Microwaves and Wireless Simplified

For a complete listing of the *Artech House Microwave Library,*
turn to the back of this book.

Microwaves and Wireless Simplified

Thomas S. Laverghetta

Artech House
Boston • London

Library of Congress Cataloging-in-Publication Data
Laverghetta, Thomas S.
Microwaves and wireless simplified / Thomas S. Laverghetta.
  p.   cm.—(The Artech House microwave library)
  Includes bibliographical references and index.
  ISBN 0-89006-908-5 (alk. paper)
  1. Microwave devices.   2. Microwave communication systems.
I. Title.   II. Series.
TK7876.L382   1997
621.381'3—dc21   97-40457
                                                             CIP

British Library Cataloguing in Publication Data
Laverghetta, Thomas S.
  Microwaves and wireless simplified.—(Artech House microwave library)
  1. Microwave communication systems   2. Wireless communication systems
  I. Title
  621.3'8415

  ISBN 0-89006-908-5

Cover and text design by Darrell Judd

© 1998 ARTECH HOUSE, INC.
685 Canton Street
Norwood, MA 02062

International Standard Book Number: 0-89006-908-5
Library of Congress Catalog Card Number: 97-40457

10 9 8 7 6 5 4 3 2 1

*To Billy Bazzy, who came up with the idea to do this book
and who has been such a good friend all these years.*

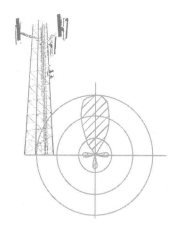

Contents

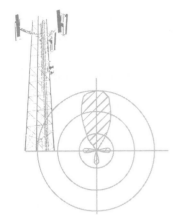

1

Introduction

I N INSTRUCTIONAL MATERIALS, the term *simple* is often tossed around loosely. Sometimes it may mean that fewer equations are used than in a typical textbook or paper. Other times, it may mean that the subject is easily understood by the author. But what does *simple* really mean? If you looked up the term in a dictionary, you might find this definition: "not complex or complicated; easily understood; intelligible, clear."

For a subject, especially a technical subject like microwaves, to be made truly simple, it must satisfy all the properties listed in the foregoing definition. That is what this book sets about to do, in down-to-earth, understandable language. And it does so with absolutely no mathematics or formulas of any kind. Now, that is really simple.

The topics of microwaves, in general, and wireless technology, in particular, generally are thought of as having a certain air of mystery to them. It is thought that to completely understand the phenomenon of high-frequency circuits it is necessary to have a large mathematical background. That is not the case. Microwaves can be understood by anyone who wants to learn about the subject. The only prerequisite is the desire to learn.

The first step in learning about microwaves is being able to define the word *microwave* in very easily understood terms, that is: "A radiowave operating in the frequency range of 500 MHz to 20 GHz that requires printed circuit components be used instead of conventional lumped components."

That definition shows that microwaves need to be treated differently from low-frequency circuits. First of all, the terms *megahertz* (MHz) and *gigahertz* (GHz) indicate frequency in cycles per second (hertz). The term *mega* (designated as 10^6) means that the signal is traveling at a certain number of million times per second. The term *giga* (designated as 10^9) means the signal is traveling at a certain number of billion times per second. Thus, you can see that the frequencies we are working with are very high.

The lumped circuits referred to in the definition are the carbon resistors, mica capacitors, and small inductors you see in your AM-FM radio or television set. The reason those components cannot be used is a phenomenon called *skin effect*, which is the concept that high frequency energy travels only on the outside skin of a conductor and does not penetrate into it any great distance. The concept of skin effect can best be understood by the following example. If you tie a string to a ball and then twirl the ball around your head at a slow speed, you will see that the ball just sort of lumbers around and stays fairly close to your head as you spin it around. If you spin it faster and faster, it begins to stretch out and be straight out away from your head and body. The force that causes that to happen is centrifugal force.

Now, let us relate the speed of the ball to frequency (slow speed is low frequency, high speed is high frequency). As the frequency gets higher, a centrifugal force also is present. The force is inductance that is set up in the transmission line simply because a current is flowing in that transmission line. This force, which we refer to as a *microwave centrifugal force*, keeps the energy from penetrating the surface of the transmission line and makes it follow a path along the *skin* of the line rather than down into the entire cross-sectional area, as in low-frequency circuits. Thus, we have a *skin effect* which determines the properties of microwave signals.

Since the high-frequency signals and transmission lines do not allow energy to penetrate very far into a conductor, it makes no sense to have round (radial) wire leads on components for microwave applications. The energy would travel only on the skin of the lead and be very inefficient. That is why you see ribbon leads or no leads with solder termination points on most microwave components. It also is why you do not see many physical components on a microwave circuit board. They are there, but they are distributed over a large, thin area and result in the same values as a lumped device that would be used at lower frequencies; hence, the term *distributed element components*. Those components are what prompt many people to look at a microwave circuit and ask, "Where are all the parts?" With these facts in mind, we can see that microwaves are high-frequency waves that require special circuit-fabrication techniques.

With a definition set forth, it now is time to get into the terminology of microwaves and wireless technology, that is, the jargon and the buzz words used by those in the microwave field.

The first term we will look at is *decibel* (dB). A decibel, which is a relative term with no units, is a ratio of two powers (or voltages). The decibel value can be positive (gain) or negative (loss). If an output power of a device (or system) is measured, an input power is measured, the ratio of the two taken, and the log of the ratio is multiplied by 10, you have a decibel value for that particular gain or loss. (When using voltages, the multiplication factor is 20.) The term decibel tells you only how much a device increases or decreases a power or voltage level. It does not tell you what that power or voltage level actually is. That is valuable in determining a system's overall gain or loss. For example, if we had a filter with a 2-dB loss, an amplifier with a 20-dB gain, an attenuator with a 6-dB loss, and another amplifier with a 12-dB gain, the overall setup (or system) would have a +24-dB gain (Figure 1.1). The value is found simply by adding the positive decibels (+32), then the negative decibels (−8), and taking the difference (+24).

Whereas decibel is a relative term, decibels referred to milliwatts (dBm) is an absolute number, that is, decibels referred to milliwatts are specific powers (milliwatts, watts, etc.). To determine decibels referred to milliwatts you need only one power. If you have a power of 10 mW (0.010W), for example, you would take that power, divide it by 1 mW, take the log of the result, and multiply it by 10 (+10 dBm, in this case).

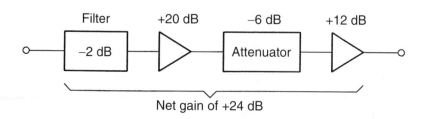

Figure 1.1 An illustration of decibels.

As can be seen, the value of +10 dBm tells you that a definite 10 mW of power are available from a source or are being read at a specific point. That differs greatly from +10 dB, which only means that there is a gain of 10 dB (gain of 10). So whenever you require absolute power readings, use decibels referred to milliwatts.

To help to understand decibels referred to milliwatts and some of the powers associated with them, see Table 1.1. The table shows five values of decibels referred to milliwatts and the powers associated with them.

The terms decibels and decibels referred to milliwatts can be used together, as illustrated in Figure 1.2. In the figure, there is an overall gain of +14 dB. You can also see that we are applying a +10-dBm signal at the input. By following the decibel and decibel-referred-to-milliwatt levels

Table 1.1
Sample Values of Decibels Referred to Milliwatts

Power	10 μW	100 μW	1 mW	10 mW	100 mW
dBm	−20	−10	0	+10	+20

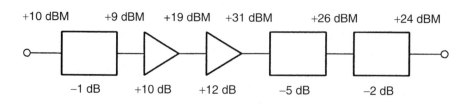

Figure 1.2 Decibels and decibels referred to milliwatts.

throughout, you can see that the output is $+24$ dBm, which is exactly 14 dBm higher than the input, just as it was when we were working only with decibels. Thus, it is shown that decibels and decibels referred to milliwatts can be used together.

A third term you should be familiar with is *characteristic impedance*. When you think of impedance, think of something in the way. A running back in football is impeded by a group of 300-pound defensive linemen; an accident on the freeway impedes the flow of traffic; and alcohol impedes one's driving skills. All these examples show some parameter in the way of normal operations. Characteristic impedance is an impedance (in ohms) that determines the flow of high-frequency energy in a system or through a transmission line. The characteristic impedance most often used in high-frequency applications is 50Ω. This value is a *dynamic impedance* in that it is not an ohmic value measured with an ohmmeter but rather an *alternating-current* (ac) impedance, which depends on the characteristics of the transmission line or component being used. You would not, for example, place an ohmmeter between the center conductor and the outer shield of a coaxial cable and measure anything but an open circuit. (A coaxial cable is a transmission line with a center conductor surrounded by a dielectric material and an outer shield. This type of transmission line is covered in detail in Chapter 3). Similarly, measuring with an ohmmeter from the conductor of a microstrip transmission to its ground plane would yield the same result. (A microstrip transmission line is a printed line on one side of a printed circuit board with a complete ground on the other. This type of transmission line also is covered in Chapter 3). This should reinforce the idea that a characteristic impedance is not a *direct-current* (dc) parameter but one that "characterizes" the system or transmission line at the frequencies with which it is designed to work.

Another point to be brought out for this parameter is that the value of characteristic impedance is the same at the input of a transmission line or device as it is 30 cm away, 1m away, or 1 km away. It is a constant that can be relied on to produce predictable results in your system.

The term *voltage standing wave ratio* (VSWR) is used to characterize many areas of microwaves. It is a number between 1.0 and infinity. The best value you can get for the VSWR is 1:1 (notice that it is expressed as a ratio), which is termed a *matched condition*. (A matched condition is one in which systems have the same impedance, so no signals are reflected back to the source of energy.) To understand the concept of a standing wave, consider a rope tied to a post. If you hold the rope in your hand and flip your wrist up and down, you see a wave going down the rope to the post. If the post and the rope were matched to each other, the wave going down the rope would be completely absorbed into the post and you would not see it again. In reality, however, the post and the rope are not matched to each other and the wave comes right back to your hand. If you could move the rope at a high enough rate, you would have one wave going down the rope and one coming back at the same time. That would result in the waves adding at some points and subtracting at others. There would be a wave on the line that was "standing still," which is where the term *standing wave* comes about.

The amplitude of a standing wave depends on how well the output is matched to the input. In high-frequency microwave applications, the standing wave ratio depends on the value of the impedance at the output of a transmission line compared to the characteristic impedance of the transmission line. It also can be shown that the standing wave ratio is a comparison of the impedance at the input of a device compared to the impedance at the output of the device that is driving it. A perfect match is indicated by no standing waves. A drastic mismatch like an open circuit or a short circuit shows a large amplitude standing wave on the transmission line or device. That would indicate a very large mismatch between devices or between the transmission line and the load that was at its output. Remember that the larger the mismatch, the larger the VSWR on the transmission line or at the input or output of a device.

A term that goes along with standing wave ratio is *return loss*. The return loss (in decibels) indicates the level of power being reflected from a device due to a mismatch. If we have a perfect match between a transmission line and a load at its output, very little, if any, power is reflected, and the difference between the input level and the reflected power is a large number of decibels. If there is a short circuit or an open circuit at the output of the transmission line, basically all the power is reflected back, and there is very little difference in decibels between the two. Thus, the return loss for a matched, or near-matched, condition is a large negative number of decibels; the value for a large mismatch is basically 0 dB. It is important to point out that the return loss is a negative number, because it is a loss. Sometimes it is difficult to understand that we have a much better match in a circuit when we have a higher value of return loss. Usually you do not want more loss in your circuits, but in this case, it is a good situation.

Another term used to describe a matched or mismatched condition in microwaves is *reflection coefficient*. The reflection coefficient is the percentage of power reflected from a mismatch at the end of a transmission line or at the input or output of a circuit. If there is a perfectly matched condition, the reflection coefficient is 0 (0%); if there is an open circuit or a short circuit at the end of a transmission line, the reflection coefficient is 1 (100%). Any mismatch condition between those two extremes is between 0 and 1. The designation for the reflection coefficient is either ρ or Γ, depending on the text you are using. This text uses ρ to designate reflection coefficient. So, if we want to have a good match for a system or a transmission line, we want to have a low reflection coefficient. If a high reflection coefficient appears, it is an indication of a large mismatch somewhere and, consequently, a high VSWR at that point.

Another term that comes into play with both microwaves and wireless applications is *wavelength*. A wavelength is the length of one cycle of a signal, as illustrated in Figure 1.3. Wavelength is designated by the symbol λ. As can be seen in Figure 1.3, one wavelength is the distance between two points that have a repeat value. If, for example, we measure 0.1V at one point on the wave, one wavelength will be where the wave

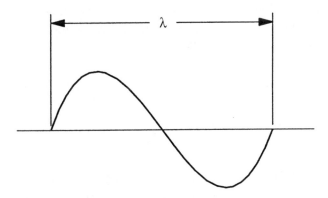

Figure 1.3 Wavelength definition.

has 0.1V again. Values used throughout high-frequency applications are $\lambda/2$ (half-wavelength) and $\lambda/4$ (quarter-wavelength). Those terms are discussed in more detail later, but for now we can say that a signal repeats itself every half-wavelength and is exactly the opposite every quarter-wavelength. A 0V signal will be zero volts every half-wavelength and maximum voltage every quarter-wavelength. The most important point to remember about wavelengths is that you should always look for points that have the same value to determine how long a wavelength is. That does not necessarily need to be where the signal is at zero, although it helps to get a good reference at those points.

A term that usually means you have a problem is *short circuit*. For high-frequency work, however, a short circuit is an intentional condition, an actual short circuit that has 0Ω if measured with a meter. A short circuit comes in handy to establish an accurate reference point along transmission lines. Care must be taken in the use of a short circuit for any application; it still is a short to dc and will short your current to ground. Remember that a short circuit is a short at 0 Hz (dc), at 1 kHz, at 10 MHz, at 20 GHz, and so on. It is always a short, so remember to correct for it.

One term that is used often but usually not defined is *wireless*, which means exactly what it says, "without wires." In a wireless communications system, there is no physical connection between the transmitter and the receiver. Although wireless technology is now a very large business, there is nothing new about the concept. Think back to your childhood walkie talkies. Nothing connected them other than air. They were (and still are) a wireless communications system. We have come a long way past that application; today, wireless *local area networks* (LANs), *personal communication systems* (PCSs), pagers, and many other systems that have no connecting wires are commonplace.

Three more terms are associated with many wireless applications: *time division multiple access* (TDMA), *frequency division multiple access* (FDMA), and *code division multiple access* (CDMA).

TDMA is a term used with many digital circuits in communications. It is a time-sharing scheme in which stations are allocated specific time slots in which to operate. Figure 1.4 shows the relationship of time and frequency for TDMA operation. It can be seen that there are specific times for each system, with *guard times* between so there is no interaction between stations. In a TDMA scheme, each channel is assigned specific

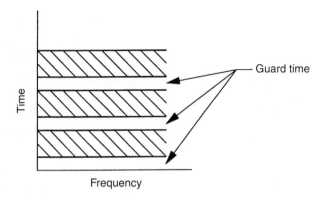

Figure 1.4 TDMA.

times to transmit and to receive. During the times not allotted to them, they cannot perform their assigned functions. That may sound serious, but remember that the times we are talking about are not 10 minutes; they are in the millisecond and microsecond range, so you will not see any interruption in your transmissions or receptions.

The next term, FDMA, is illustrated in Figure 1.5. Using the same time and frequency references as in Figure 1.4, Figure 1.5 shows that each station in the FDMA case is on all the time but is assigned certain frequencies in which to operate. There also are spaces between stations

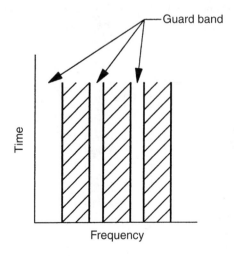

Figure 1.5 FDMA.

in this scheme, called *guard bands*, which serve the same purpose as the guard times in TDMA. FDMA is the method with which most people are familiar (although they may not realize it), because it is used for AM and FM radio and television. Each station, or channel, is assigned a specific frequency on which to transmit. The stations are on all the time at their assigned frequencies. There also are bands between stations so you do not get an easy listening radio station moving in on a rock station or a television sitcom interfering with the evening news.

Finally, CDMA is the scheme used for spread spectrum secure communications systems. Figure 1.6 shows the same time and frequency references as were used in Figures 1.4 and 1.5, but this time no specific time or frequency is allocated. CDMA uses *chips*, which are specific times and frequencies. That is where the concept of secure communications comes into effect. Usually, a pseudorandom code is established at the transmitter and is received only by those receivers that have the same code, so they can receive the signal and demodulate it. This is an important part of the cellular telephone operation, because it makes the telephones, and consequently your conversations, secure, something not available in the first cellular telephones. In the early days of cellular telephone operation, anyone with a regular scanner could pick up and listen in on a conversation. With a CDMA approach, conversations are secure.

Using the basic terms presented in this chapter, managers, marketing personnel, and sales personnel should be able to converse with micro-

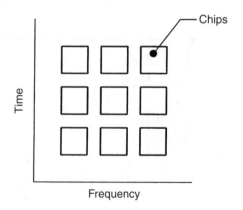

Figure 1.6   CDMA.

wave personnel to establish requirements for particular applications. Other terms come up throughout the text, and they will be defined and explained as they appear. There is also a glossary in the back of the text. Now it is time to get into the actual microwave and wireless applications and operations.

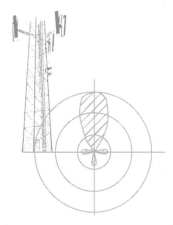

2

Microwave Applications

N OW IT IS TIME to get into some common applications of microwave and wireless systems. When most people hear the term microwave, they immediately think of microwave ovens. That is natural and perfectly all right, since microwave ovens operate at 2.45 GHz, which is in the microwave and wireless frequency band. Also, a microwave oven is a small variation of radar, an application covered in this chapter. So, you can see how natural the association really is to someone who does not have a background in high frequencies.

The microwave oven is a device with a high-power tube (magnetron) that sends energy into food to be prepared. It does so by heating the moisture inside the food. That is why the food cooks from the inside to the outside. If you ever happened to put your finger on the center conductor of a coaxial cable with microwave energy propagating along it, you would notice a white mark on your finger. The mark would be below the skin, and the skin would not be broken. The microwave energy would use the moisture in your body and heat it to begin a cooking process below the skin. That is what happens when you put food into a microwave oven and turn it on. (If you look for the microwave oven on a microwave frequency chart, you will not find it designated as such. What you will find is a section called "microwave heating.")

Let us take a look at the electromagnetic spectrum and, in particular, the microwave spectrum, to further understand what frequencies we are talking about when we discuss applications or other aspects of microwaves. Figure 2.1 is a drawing of the electromagnetic spectrum. Notice that it covers a wide range of frequencies, from a few megahertz to the visible light spectrum and higher. You can see from this spectrum representation that there are many applications for *radio frequency* (RF) and microwave signals. This chart shows only a few of them. (Notice the absence of microwave ovens and no reference to microwave heating. That is because this text concentrates on the commercial applications of microwaves that are related to wireless technology. The microwave oven certainly is not wireless, by any stretch of the imagination).

Some of the more recognizable applications shown in Figure 2.1 are AM and FM broadcast bands for radio, television channels, cellular

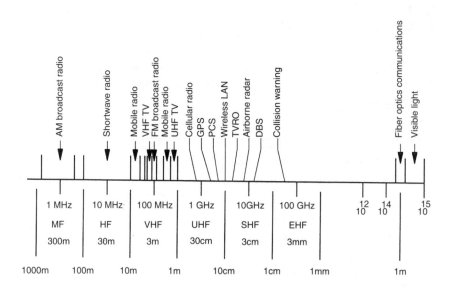

Figure 2.1 The electromagnetic spectrum.

phones, *global positioning systems* (GPSs), PCSs, and *direct broadcast satellites* (DBSs). Each of these applications has a different frequency of operation. That is, they operate in an FDMA mode, for the most part. You will recall that, when in the FDMA mode, a system operates over a specific band of frequencies all the time. There is no time gap planned, and no time sharing of stations or channels. They are there all the time under normal operating conditions. Some applications are TDMA devices, in which there is time sharing, and those will be pointed out as we get to them.

The applications presented in this chapter are divided into three sections: radar; telephones and telephone systems; and telecommunications, specifically wireless. Each type of application is presented and discussed in enough detail to give the reader a general knowledge of each topic. Terms are presented and defined, and examples of each application are presented.

2.1 Radar

Until recently, whenever microwaves were mentioned, most people thought either of the microwave oven (as we have said) or of radar. To some extent, that perception has changed. Many radar applications with new variations are being used everyday, for example, in the areas as medicine and collision avoidance systems, to name only two. Actually, the term *radar* has taken on different meanings as new and improved applications are found.

The term radar was originally short for radio detection and ranging. With the changing technology, the definition has been altered slightly to the following: "an electromagnetic device for detecting the presence and location of objects." That is really a much more valid definition for radar in the modern world. The basic principle behind radar is that of a transmitter sending out a very short duration pulse at a high power level. The pulse is controlled by a pulse-forming network and begins the time sequence when it is transmitted. The pulse strikes an object or target and reflects the energy back to the radar receiver. The time it takes for the pulse to be transmitted, bounce off an object, and be received determines the distance that object is away from the radar antenna. The concept is illustrated in Figure 2.2.

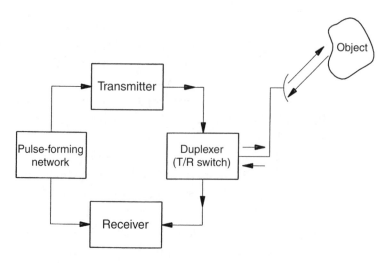

Figure 2.2 A basic radar system.

An additional block in Figure 2.2 is the duplexer, or *transmit/receive* (T/R) switch. The duplexer is a circuit that switches the antenna from the transmitter to the receiver at the proper time so the signal can be transmitted to perform its tasks without destroying the receiver in the process. At the same time, the switch allows the very low level signal coming back from a reflection to be sent to the receiver and not back into the transmitter. The duplexer can be a physical switch or a series of transmission lines that performs the switching functions. Such a switch is important for proper operation of the radar system, because it protects the system's receiver.

To further understand the concepts and operations of radar, it is necessary to understand some of the terminology that is used to refer to the parameters of a system. The first term we will look at is *continuous wave* (CW). This term, which is illustrated in Figure 2.3(a), refers to a signal that is on continuously. As can be seen in the figure, there is no time that the signal is off. Basically, all signal generators use CW to supply signals to individual systems. This type of signal is what is being generated in a lab when systems or components are being tested, or when you want to test your television amplifier to make sure it is still working. It is ideal for systems that need to have power to them at all times.

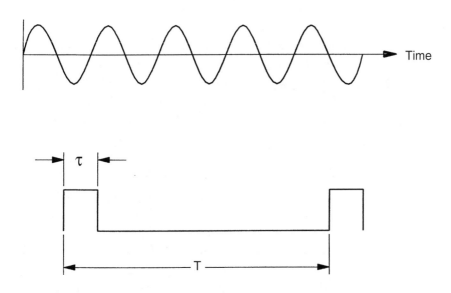

Figure 2.3 Radar Signals.

The second type of transmission, shown in Figure 2.3(b), is the heart of a radar system and what actually makes the whole concept work, the *pulse*. A pulse type of signal supplies power for only very short amounts of time, which allows for some very high powers. That usually is not possible with CW systems, because it would take a lot to have megawatt powers that were on at all times. Also, if the power is on all the time, there is a potential problem with the components in the system being able to dissipate all that power. If, however, the signal is on only 5 to 10% of the time or less, it is possible to obtain higher powers over a short duration. That is possible because either you have a certain amount of energy available, spread it over a long period of time, and have it be a small amplitude, or you take the same amount of energy, have it on for a short period of time, and have it be a much higher amplitude. That is the idea behind the short-duration, high-amplitude pulse systems.

To understand a pulse system, you must be familiar with some terms. The first is the *pulse width*, which is designated as τ, tells how long the signal is on (in seconds, milliseconds, microseconds, etc.). The pulse width is an important term to know, since it determines the actual operation of the pulsed system. How long the device actually is on helps to determine many parameters for the entire system. A second term that goes along with the pulse width is the *pulse repetition rate* (PRR) or, as it is called in some texts, the *pulse repetition frequency* (PRF). The PRR, designated as "T" in Figure 2.3(b), tells you the amount of time between pulses, that is, how often the pulses occur in your system.

With the two terms for a pulse defined, we now put those definitions together to get a term that is used in all pulse applications, *duty cycle*. The duty cycle is the ratio of the pulse width, τ, to the pulse repetition rate, T, that is, τ/T. The duty cycle is the 5%, 6%, 7%, 10%, or whatever percentage of time that the signal is actually present. Looking at it another way, it is the time that the signal is actually doing something, or is on duty, compared to the total amount of time available (between pulses). This parameter is a vital one in the characterization of any radar system. It is the one that tells an operator or a designer what the radar, or pulsed system, actually has available to do the tasks necessary.

Another term that needs to be addressed is *peak power*, which is the amount of power present at the top of the pulse. In Figure 2.3(b), the peak power is the amplitude of the pulse over the duration, τ. Peak power usually is quite high, but it is present for only a short period of time. You will see many components characterized with both a peak power specifications and CW power specifications. That is so you can use them in either application and not have to worry about the power that is being applied.

When you have a pulse system, you also will be concerned with *average power*. Average power is defined as the peak power multiplied by the duty cycle. Look at Figure 2.3 again to see how that is the case. The power is available for the period of time that the pulse is on and is the peak power. The next pulse that comes along also contributes to the average power of the system and is also considered. Thus, the pulse repetition rate and the pulse width are needed. That, you will recall, is the duty cycle. So, the result is that the average power is the peak power times the duty cycle.

The applications and functions of radar systems can be categorized as follows: search and warning, tracking and measurement, imaging (identification), and control and communications.

2.1.1 Tracking and measurement applications

The tracking and measurement function of radar is the one most people think of when they hear the word *radar*. It is the detection of a target (an airplane in the air, a land mass on a radar scope, etc.) Such targets usually are struck many times by the signal because of the many scans by the antenna. A typical area where you would notice a radar system is at an airport. It is especially noticeable at smaller airports, where the antennas are much more visible and can be seen rotating. Larger airports have their antennas, many of them protected by domes, in much more remote areas.

Another area where this type of radar system is visible is at docks. The freighters, tankers, and cruise ships all have radar systems on them with rotating antennas for navigation purposes. Also, most luxury crafts also have their own radar systems with rotating antennas.

Measurement and tracking radars lock on to a target and track it for a certain distance or for a certain time period. Military applications of radar systems are for gun control and missile guidance. Imagine how difficult it would be to aim a ship's guns or missiles in the desert without radar systems. It could be done, of course, but the accuracy would be practically nonexistent. Many more international incidents would occur without this type of guidance system. Figure 2.4 shows a typical tracking radar.

2.1.2 Imaging applications

Imaging radar operates by taking a single target from a large field of objects and forming an image that is two- or three-dimensional in nature, usually in azimuth and range coordinates. This type of system analyzes mechanical systems for stress and is used with very low power transmitters for some medical applications (see Figure 2.5). Such a scheme takes a tumor, for example, and makes a three-dimensional picture of it to give doctors a much better picture of what they are dealing with. Figure 2.6 shows imaging radar displays.

2.1.3 Doppler radar

We now look at a special-condition radar, Doppler. Doppler radar was originally conceived for areas with mountainous terrain, where it was difficult to detect moving targets. Before the advent of Doppler radar, it

Figure 2.4   Tracking radar.

Figure 2.5   Imaging radar.

Figure 2.6 Imaging radar display.

was easy for an aircraft to slip into a mountainous region and proceed virtually undetected to a target. Conventional radar would not indicate a moving target, just a target, which could be an actual airplane or one of the mountains.

You probably have encountered the Doppler effect on many occasions. For example, when you stand at a train crossing and an incoming train blows its whistle, you notice a change in the pitch of the sound as the train approaches and then passes. If you can measure the change in pitch, you can identify a target and tell its velocity. That is the principle behind police "speed traps," which use Doppler radar systems and are very accurate. Such systems are difficult to detect in time for speeding drivers to slow down. Usually by the time you have detected it with a radar detector, it is too late; the radar system already has recorded your speed.

Doppler systems concentrate on moving targets. The signal is sent from the radar transmitter at a certain frequency. When the signal strikes the target, it reflects it back to the receiver; the frequency that comes back to the receiver determines the speed of the target. If the object (target) is moving toward the receiver, the frequency appears to increase. Similarly, if the object is moving away from the receiver, the frequency appears to decrease. By measuring the change, certain parameters can be determined about the detected object (range, speed, etc.). Systems like these have many applications on manufacturing assembly lines, in which the position and the speed of a product coming down the line must be determined so certain operations are performed at specific times and at specific locations.

The most common type of Doppler radar is the police radar. A basic diagram of this type of radar is shown in Figure 2.7. A transmitter/receiver block in the police car or on the side of the road sends out the signal, which strikes the moving automobile and returns to the receiver. The key term here is *moving;* Doppler systems cannot detect stationary objects. After detecting a moving vehicle, the Doppler system displays the speed of the car on a screen. Figure 2.8 is a picture of a police radar device.

Another application of Doppler radar is the Doppler speedometer. The antenna is under the vehicle, and the reflections return to the receiver

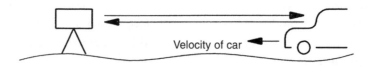

Figure 2.7   Police Doppler radar

Figure 2.8 Police radar unit.

and indicate the distance moved and the time elapsed, that is, the velocity of the vehicle. When the vehicle stops, the speedometer indicates zero. When the vehicle moves, the speedometer indicates its relative velocity with respect to the ground, which is exactly what a speedometer is supposed to do. Figure 2.9 shows the concept of such a device. When the car is being driven on a smooth highway, the reflections are very accurate and give an accurate velocity for the vehicle. Even on a country road, a Doppler speedometer cancels out most of the extra reflections that may occur and still gives an accurate reading.

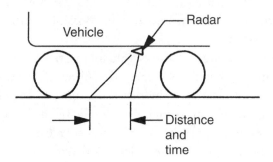

Figure 2.9 Doppler speedometer.

An application of Doppler radar that has been around for many years is that of a collision avoidance system. This application seems to have had a difficult time finding acceptance within the automotive community and the general public. For some reason, it has taken many years to have such a system accepted by anyone. Even today, there are people who do not trust such a system to really provide collision avoidance. There are many variations of this type of system, but the basic principle is as follows. The system sends out a radar signal and monitors the distance between your car and the vehicle in front of you. If the distance gets smaller, the system causes your vehicle to slow down, thus avoiding a collision. The change in distance results in a proportional change in frequency.

A variation of the collision avoidance scheme is being tested to detect objects behind a vehicle when it is backing up. Such a system will detect tricycles, toys, and, most important, children who may be playing in a driveway or walking behind parked cars. The driver is warned in time to stop before striking an object or a person behind the car. This tremendous safety feature is also an excellent device for construction equipment and large vehicles like buses and trucks. Figure 2.10 shows another common object detection system.

A more recent application of the Doppler effect is weather radar to predict the paths of severe weather, such as tornadoes and hurricanes. Doppler weather systems can spot a storm, track it, and allow weather bureaus to warn and evacuate people before the storm arrives. The Doppler system indicates the motion of the target (e.g., storm clouds). We all have seen the displays on television of hurricanes as they develop in the Atlantic or Pacific oceans and approach the mainland. It is interesting to look at the intensities of the storms and watch them develop into full-scale hurricanes. That would not be possible without the use of Doppler radar systems.

This same type of system also is valuable in the Midwest portion of the United States. Particularly during the summer, a large number of tornadoes occur in this area. Winds inside a tornado have been clocked upward to 300 mph. The movement is detected by the Doppler radar and indicated on a screen. Appropriate warnings then can be sent out to the areas that will be affected by the storm, and lives can be saved. There probably will not be much hope for property in the way, but that can be rebuilt.

Figure 2.10 Object detection system.

These are but a few of the applications of radar. We have come a long way from the radar systems designed to detect enemy planes and ships during World War II. Radar systems have become sophisticated and are very much in use for both civilian and commercial purposes. Think what an "adventure" it would be to fly on a commercial airliner without the use of any radar (weather radar, radar altimeters, navigation radar, etc.). Not a particularly comforting thought, is it? The applications of radar reach into every life in this country and the world, probably much more than any of us really realizes.

2.2 Telephones and telephone systems

Stop and think what your average day would be like without a telephone. It might be a lot quieter, but you probably would not get much accomplished, and you would spend a lot more money on gasoline and airplane tickets to get some things accomplished. Modify that thought a bit more and allow yourself a telephone but take away your fax machine along with every other fax machine in your company. Now, awaken from this nightmare because you do have all these technological marvels at your fingertips. There are people who abuse these devices, but telephones and fax machines make your life much more productive and much easier than it may have been in the past. Twenty-five to 30 years ago, you would have been laughed out of the room if you had suggested that you could walk around the room with a telephone, make a phone call from your car, or have teenagers with telephones walking around a mall. Today, it is second nature to have those facilities available to you. Now, most people ask for your fax number or your e-mail address as much as for your telephone number. So, things change, and the telephone is a large part of those changes.

The basic telephone has been written about many times and in many different ways. To bring the telephone into the realm of this text, we have to mention only one term: cellular telephone. The cellular telephone is the modern-day version of mobile communications. Mobile telephones actually originated in the late 1940s but never found wide use because of the high cost and the limited frequency allocation. In the 1970s, this last restriction was removed when the 800 to 900 MHz band was allocated for mobile communications. Also, as the technology has advanced, the cost of a mobile communications system (cellular telephone) has come down considerably.

The cellular concept can best be pictured as a group of automatically switched relay stations. A populated area is divided into many small regions, called *cells*. The cells are linked to a central location, called a *mobile telephone switching office* (MTSO), which coordinates all incoming calls. Along with coordinating calls between cell sites, the MTSO also generates time and billing information. A diagram of a cellular system is shown in Figure 2.11. This simple diagram presents the basic blocks of a cellular system. Notice the cells at the left of the diagram. Each cell has a transmitter/receiver combination in it that is for a certain section of an area. The main block, the MTSO, is the control area for the cellular telephone system. It is the unit that connects the caller to the party the caller is trying to contact. If the caller is moving (e.g., in a car), the MTSO senses the level of the signals being used and automatically switches the call to the appropriate cell so the transmission is completed with the best clarity possible. The central office in Figure 2.11 provides the same functions as the central office in a conventional telephone system, that is, it provides a connection between one phone and another.

The cell site is actually a special transmitter/receiver combination. Because it covers only a small geographical area, the unit is relatively low power. That allows other cells to operate on the same frequency, since the power is low enough that no interference occurs. This feature is important since the many cells in an area would interfere with each other if it not for the low power requirements placed on each cell. Thus, many cells can exist in a geographical area, all operating at the same frequency and coexisting very nicely. Figure 2.12 shows a cell site for cellular telephones.

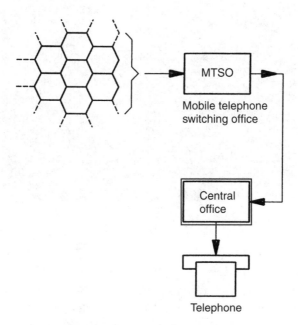

Figure 2.11 Cellular telephone system.

Figure 2.12   Cell site for cellular system.

The 800 to 900 MHz frequency band that has been allocated for cellular telephone service ranges from 825 to 845 MHz and 870 to 890 MHz. For the cellular phone, the lower end (825 to 845 MHz) is used for transmitting, while the upper end (870 to 890 MHz) is used for receiving. At the base units (cell sites), the frequencies are reversed. This approach is logical, because a phone's transmitter is the cell's receiver and vice versa. Within the assigned bands, 666 separate channels are assigned for voice and control, 333 in each band. The bandwidth for each channel is 30 kHz.

A person making a cellular telephone call enters a local 7-digit number or a long-distance 10-digit. The caller then presses the send button, which sends data to a channel. From the cell site, the data are forwarded to the MTSO with the cell site's identification number. Once the MTSO detects that the cellular phone is on the proper designated channel, the call is sent to the central office and then to the "callee's" phone. This sounds like a time-consuming order, but it is accomplished in a very short period of time.

When a cellular phone's signal strength decreases because of the distance that has been traveled, the MTSO searches through the cells to find the one with maximum strength and automatically switches the conversation. This process is called a *handoff* and is a process that the user never sees or is even aware of. You probably have encountered handoffs many times during cellular phone conversations and never even knew it was taking place. This is one of the truly outstanding features that make cellular telephones so popular and in demand. It would be very annoying during a phone conversation to have the signal fade and eventually disappear. A lot of business would not get done, and a lot of marriages might not survive if that were the case.

Most cellular telephones are used in a local area where the phone is originally registered. When it is necessary to operate the phone outside that selected area, the system must incorporate a scheme called *roaming*. Roaming is possible only if the area you are in has cellular service and agreement has been reached between telephone companies and their users. Many areas of the country have roaming capability, thus extending the range of cellular systems greatly. Roaming is a handy feature for those who do a lot of traveling to different parts of the country. Cellular telephone users who plan to travel outside their local areas should check with their carriers to make sure all the features they need are available. A little time spent checking things out before you leave on a trip decreases the likelihood of frustration later on when you try to make a call from your rental car that will not go through. Figure 2.13 shows a cellular telephone.

Figure 2.13   Cell phone.

A further advance in the area of mobile communications in the RF and microwave area is the *personal communications network* (PCN). Figure 2.14 shows PCN equipment. A PCN is designed to operate independently as well as interface with the standard telephone system. PCNs operate in the 1.9 GHz frequency band. The transmitter/receiver units are designed to be very small and have low-power outputs. That means more base stations are required for PCNs than for the existing cellular telephone systems. A *personal identification number* (PIN) as a part of a PCN eliminates the need for separate numbers at different locations (home, work, etc.).

The telephone business has come a long way from the party line, which assigned customers a specific number of rings. It also has come a long way from the hand-cranked wall telephone, which required that an operator place calls for you. The concepts, however, have not changed, only the technology that accomplishes them. Customers no longer have to count the number of rings before they answer the phone; the coded rings are taken care of automatically. Operators have been replaced by control stations that do the same task, only much faster and, in some cases, more accurately. All that is done with RF and microwave technology.

Figure 2.14 PCN ITA equipment. *(Photograph courtesy of Assessment Services Limited.)*

2.3 Telecommunications

Telecommunications is a term that is used to denote many things. Sometimes it is used by people who really do not know what area of communications they mean. At other times, it is an impressive term to use. But what does it really mean? The prefix, *tele*, means far off, distant, remote. Thus, the word telecommunications refers to a process of communications over a long distance. For our purposes, we mean electronic communications (technically, smoke signals and homing pigeons also could be categorized as modes of telecommunications.) Older textbooks on telecommunications describe them as communications over a wire. To some extent, that is still true today, but a great deal of communications are carried over optical fibers and by means of RF and microwave signals.

This text concentrates on the RF and microwave telecommunications. We look at telecommunications as being the transmission and reception of information over various distances by means of a microwave signal. In particular, we deal with applications that are termed *wireless,* a term that is becoming more and more commonplace and that might well be the buzz word of the 1990s as those technologies develop and thrive.

If you stop and think about it, wireless systems have been available for many years. As was mentioned in Chapter 1, toy walkie talkies are wireless because they use antennas and no interconnecting wires between transmitter and receiver. Truckers have used *citizens band* (CB) radios for many years to communicate with one another and with motorists. So, the idea of wireless communications is not new by any means. It just has come of age and is following the new technology and the public's demands.

This section on telecommunications looks at some of the wireless applications that use RF and microwave signals and transmit them through the air rather than propagate them along a coaxial transmission line or an optical fiber. The topics we cover are cordless telephones, personal communication systems, and wireless LANs.

Just about everyone is familiar with cordless telephones (see Figure 2.15). The idea of a telephone as a device that is wired into the wall has seen its day. In a cordless telephone system, the base is connected to a conventional telephone line and contains an RF transmitter/receiver that sends and receives signals to and from the handset, which is not connected to the base and which the user can carry around. The handset is battery operated and also contains an RF transmitter/receiver so it can communicate with the base station. The system operates very well (static-free operation and a clear, understandable signal) when the hand-

Figure 2.15 Cordless phone.

set is within 50 to 500 ft of the base. Of course, some units do not even begin to approach those numbers, while others do it very nicely.

One problem with some cordless telephones is that they are not secure. It is possible for an unauthorized third party to listen in on conversations taking place on analog-type cellular phones. Digital-type cellular phones address that disadvantage, and their claim to fame is that they offer secure communications. Such security is accomplished by a technology developed for military radios called *spread spectrum*. The transmission jumps (or hops) between several frequencies, making the frequency being used difficult to detect.

Another RF and microwave wireless communications system is the *personal communications system* (PCS). The idea behind a PCS is the ability to reach people who are away from their desk, office, or wherever they normally could be found. PCSs transmit on frequencies of 1850 to 1910 MHz and receive on 1930 to 1990 MHz. A PCS allows the user to be contacted in virtually any location. A commonplace example of a PCS is the pager. How many people have you seen who have pagers attached to their belts, in their purses, or in their briefcases? The number is probably uncountable. That is because pagers have become absolutely necessary in this fast-paced work world.

A pager is similar to a miniature cellular telephone (see Figure 2.16). The major difference is that a cellular telephone transmits and receives in both directions. The pager uses a one-way path; a message is transmitted to the pager, where it is displayed for the user to act on. Usually that involves a telephone call to the number listed on the display. Other times, a message is displayed, like "Happy Birthday" or "Meet you at the corner of State and 5th Street at 4 p.m.," so no reply is required. The pager has become another very important part of life for many people.

The last topic to be covered in this telecommunications section is that of wireless LANs. The term LAN is used for many applications in this computer age we are in and will be in for some time to come. It takes into account such areas as bank transactions, credit card checks at stores, airline reservations, and hotel reservations, and on and on it goes. Many of these applications are carried out using coaxial cable connections or optical fibers to get information from one point to another quickly. Other areas use wireless transmissions to cover great distances and get informa-

Figure 2.16　Pager.

tion to many more people. Wireless applications, of course, involve modulating an RF or microwave carrier with the information to be transmitted and sending it out through the air to get it to the proper location(s).

To explain further just what a LAN is, we can say that it is a collection of personal computers or other workstations within a limited distance of one another and connected by a type of communications facility. The communications facility for wireless applications is either an *infrared* (IR) or a radio-based type of facility.

IR systems are like the remote control units used for television sets. The main problem with the IR type of system is that it must operate in a line-of-sight mode. That is, the transmitter and the receiver must be in a straight line. For example, when you try to change the channel on your television from an oblique angle in the room, you can click all you want, but the channel does not change. That is because you are not lined up with the IR sensor in the TV itself.

A radio-based system operates in the microwave range and the *ultra-high frequency* (UHF) spectrum (300 MHz to 3 GHz) to transmit signals between transmitters and receivers. LANs operate at 2400 to 2483 MHz, while the larger *wide area networks* (WANs) receive at 935 to 941 MHz and transmit at 896 to 902 MHz.

Many different systems are used for wireless LANs. Some systems use a master control transceiver that transmits and receives from a group of workstations. Other systems use Ethernet-type schemes or token rings, in which each node has its own wireless LAN card and plug-in antenna. (A token is a code that is passed around the ring of stations, which are connected in a continuous pattern. With such an arrangement, the station that wants to transmit must possess the token, or code.)

Wireless LAN technology has progressed to the point where the *Institute of Electrical and Electronics Engineers* (IEEE) has established a standard (802.11) that characterizes the requirements for such a LAN.

2.4 Summary

There are many applications in the RF and microwave spectrum for both military and commercial areas. This chapter discussed only a few of those applications. Specialized texts present a great deal more information on these topics; this book is designed to be a general overview of topics that allows the reader to go on to the specialized texts, if he or she so desires.

This chapter discussed radar, telephones and telephone systems, and telecommunications. For each topic, we concentrated on specific applications (Doppler radar, cellular telephones, pagers, and wireless LANs). This coverage should give the reader a general idea of what can be done in the RF and microwave areas concerning everyday applications of a highly developed technology.

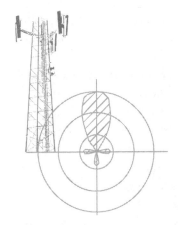

3

Transmission Lines

THE GENERAL DEFINITION of a transmission line is a device that transfers energy from one point to another. If that is the case, a clip lead used in a dc circuit would satisfy that definition. For dc and low-frequency applications (e.g., audio), this is a good definition of a transmission line. Usually it is not called a transmission line but simply a wire. Regardless of the terminology, it is a transmission line for such applications. Obviously, a clip lead would not work very well for RF and microwave applications. Chapter 1 discussed how the skin effect causes all sorts of problems with high energy traveling down a transmission line if that line is not designed exactly right. If we used a clip lead as a transmission line for a 1.9-GHz application, for example, there would be very little energy at the output of a clip lead because of a number of factors. Most of those factors are due to the skin effect in the line itself. There would be power lost in the transmission line and power lost because of radiation from the line. Thus, we have to change our definition of a transmission line. To take into account all the possible arrangements for high-frequency transmission lines, we now make the definition of a transmission line as "a device used to transfer energy from one point to another efficiently."

This definition makes much more sense for all applications, including RF and microwaves. By efficiently, we mean a device with a minimum amount of loss through it and reflections from it. It should be as close to a perfect match (VSWR = 1:1) as possible. That is important at RF and microwave frequencies, because as the frequencies get higher and higher any energy lost in a transmission line or component is much more difficult and costly to get back. That is why a low-loss, efficient transmission line is so important in microwave and RF applications.

An equivalent circuit for a section of transmission line is shown in Figure 3.1. The representation is of only a single section of the transmission line, not the entire line. (The dashed lines on both sides of the figure indicate that there is more transmission line than is shown.) There are four parameters to look at: inductance (L), resistance (R), capacitance (C), and conductance (G). There is also the dielectric constant, ε, of the material in the transmission line.

Values for each of the four parameters are expressed in their appropriate units per unit length. The values are expressed per unit length because, as you will recall, the equivalent circuit shown in Figure 3.1 was of only a section of the transmission line. That section length is the unit length to which we are now referring. The length can be any unit appropriate for your particular application (feet, meters, centimeters, etc.). The value simply indicates what each parameter is for that length of line. Now let us look at each parameter to learn more about the makeup and the operation of the transmission line.

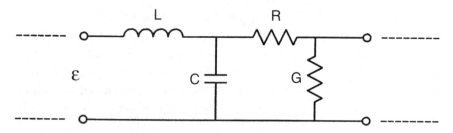

Figure 3.1   Equivalent circuit of a transmission line.

The inductance (microhenries/unit length) in a transmission line comes about because a current is flowing in a metallic conductor. In basic motor theory, inductance is one of the first concepts that must be grasped. With an ac (which is what RF and microwave signals are) flowing through a conductor, a magnetic field is set up by that current. The field reaches its maximum at the maximum amplitude of the current flowing. When the cycle reverses and begins to go in the opposite direction, the magnetic field collapses and generates a current in the opposite direction. The expansion and collapse of fields produce an inductance in the wire that is carrying the current. That is how electric motors operate. Without that phenomenon, there would be no electric motors and fields would not be set up in RF and microwave circuits on their transmission lines. The arrangement of fields sets up an inductance on the transmission line that can be characterized as an inductance per unit length, as previously stated. The value should be relatively low for proper operation of any RF or microwave transmission line. Inductive reactance, which is the ohmic result of inductance, increases with frequency and can cause problems for high-frequency circuits. Thus, the inductance should be kept low for proper operation.

The resistance (ohms/unit length) shown in Figure 3.1 also is associated with the metallic conductor and current flow. Any time you have a current flowing through a metallic conductor, there will be a loss because there is a certain resistance in that conductor. This concept follows Ohm's law, which says that there is a voltage drop across a resistance when a current flows through that resistance. The loss in the conductor of a transmission line is caused by a current flowing through the resistance of that conductor. Thus, a voltage drop occurs, and there is an additional loss of signal from that of the input to the transmission line.

The last two parameters are associated with the dielectric used in the transmission line. The first parameter is capacitance (farads/unit length). To build a capacitor, you have to have two plates of a certain area separated by a certain distance with a dielectric between them. If you look at Figure 3.1 again, you will see the two plates, which are the upper and lower conductors of the transmission line. One plate is the center conductor of the transmission line and the other plate is the ground, or shield. These plates have a certain area to them, and they are separated by a certain distance, which is the dielectric (ε) in between. So it can be seen that we have set up a very good capacitor with the two conductors and the dielectric between them. That is the capacitance of the transmission line over a certain length, and it should be kept to a minimum, just like the inductance. The capacitive reactance, which is a result of the line capacitance, decreases with an increase in frequency and causes the signals you are trying to propagate to be shorted to ground at certain frequencies. So, the capacitance should be minimized on transmission lines.

Conductance (siemens/unit length) is the amount of leakage through the dielectric. There is always a certain amount of conductance, because there is no such thing as a perfect dielectric. That means a certain amount of energy going down the transmission line appears at the other conductor. It usually is a very small quantity, because many dielectrics are very good insulators in applications such as a transmission line.

To further clarify the important characteristics of transmission lines, let us assume we have a perfect transmission line, that is, $R = 0$ and $G = \infty$. Those values indicate that the resistance of the conductor is so low we can ignore it and that the dielectric is perfect (no leakage). The result of this assumption is shown in Figure 3.2, the circuit diagram for a lowpass filter, which is a component that passes everything below a certain frequency (its cutoff frequency) and attenuates everything above that frequency. (Lowpass filters are covered in detail in Chapter 4.) With this characteristic, we are saying that a transmission line has some frequency, depending on its individual characteristics, above which you do not use the transmission line. This very important characteristic of transmission lines should be noted and observed in order to have a circuit or system work properly. Operation above the cutoff frequency results in much higher losses in the transmission line than does operation in the proper frequency range. Enough problems come up in RF and microwave circuits without having to deal with the wrong transmission line being used.

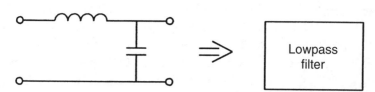

Figure 3.2   Lossless transmission line.

A transmission line, when used to connect circuits, must be basically transparent, or invisible. That is, you know the line is there connecting the circuits but its effect on the circuit is minimal, which is the ideal situation. Some terms associated with transmission lines must be explained here: VSWR, reflection coefficient, and return loss. (These terms were presented in Chapter 1 as general definitions. They are presented again, in much more detail, to relate their importance to transmission lines and their operation.)

To understand VSWR, visualize a rope tied to a door. If you take the untied end in your hand and flip it up and down once, you will see a "wave" going down the rope to the door. If the door and the rope were completely compatible, the wave would disappear into the door. But, as we all know, the wave does not disappear, it comes right back to you. (As a matter of fact, if you do not pay attention to what you are doing, the rope will flip right out of your hand when the wave returns.) That is the same condition for an RF signal or a microwave signal as it travels down a transmission line. If the wave goes down the line and the impedance of the load is exactly equal to the characteristic impedance of the transmission line (usually 50), the signal is completely absorbed into the load and no energy is returned. This is a perfectly matched condition (i.e., VSWR = 1:1).

If the load impedance is not equal to the characteristic impedance of the transmission line, a certain portion of the signal sent down the line is reflected back. That signal adds to the incoming signal in some cases and subtracts from it in other cases. The result of the addition and subtraction is a wave that "stands still" on the line, thus the term *voltage standing wave ratio*. It is called a ratio because it is the comparison of the maximum amplitude of the voltage in the standing wave to the minimum amplitude of that same voltage. Since the comparison is between voltages, the VSWR value has no units and it is a ratio compared to 1.

The best value you can have for a VSWR is a 1.0:1 ratio. That means the load and the transmission line are the same impedance. The other extremes are an open circuit and a short circuit. With an open circuit, virtually all the energy sent down the line is reflected back because it has no other place to go. Theoretically, all the energy is reflected. Practically, a small amount is radiated out the open end of the line. A short circuit also reflects all the energy sent down the line. That is because no voltage can be developed across a short circuit. So there is no signal to be sent to a *load*. Recall from Chapter 1 that a short circuit is one of the best friends we have in RF and microwave testing. It is a way of knowing the exact location of a component at the end of a transmission line and sets up a reliable reference point for all the measurements that need to be taken. Thus, a short circuit is good for testing and represents an extreme case when we are looking at the VSWR of a device or circuit. The VSWR for both an open circuit and a short circuit is ∞. That, of course, is the largest theoretical value you can get. In actual practice, you get VSWRs of 10:1, 20:1, or 25:1 as indications that you have something like an open- or short-circuit condition. Obviously, you will never see a VSWR of infinity because that is a theoretical value used only for calculations and explanations. A VSWR like the ones presented here is about as close to those conditions as you can get.

Everything else between an open- or short-circuit condition and a perfect match is a number that is a ratio compared to 1. Those numbers will be between 1:1 and ∞, depending on the degree of mismatch in the system.

It is difficult, if not impossible, to get a perfect 1:1 match in a practical circuit. If you can achieve a 1.5:1 or a 2:1 match, you are doing pretty well. Some devices get down to 1.2:1 or slightly lower, but that is about as low as you will see in everyday applications. Connectors will exhibit VSWRs on the order of 1.05:1, but that is absolutely necessary, since the connectors used on a transmission line must be excellent in performance and not take away from what the circuit is attempting to do.

Another term used to describe the match on a transmission line is *reflection coefficient*, which is the percentage of signal reflected back from the mismatch. The reflection coefficient is expressed in a couple of ways. You may see it as ρ or as Γ. Either symbol can be used, but make sure you know what the particular author is using before investigating a transmission line, because the same symbols also are used for VSWR.

Because the reflection coefficient involves reflections on a transmission line, it is directly related to the VSWR. To determine the reflection coefficient when you know the VSWR, divide (VSWR − 1) by (VSWR + 1). That gives the percentage of signal reflected back by the load impedance. For a perfect match, no signal is reflected by the load, so the reflection coefficient for that case will be zero, that is, 0% is reflected back. For the open and short circuits, all of the signal is reflected back, so the reflection coefficient in those cases is 1, that is, 100%. In other words, for the perfect match, 0% is coming back, and for open and short circuits, 100% of the signal is being reflected back. (This discussion concentrates only on the magnitude of the reflection coefficient. The reflection coefficient also has a phase angle associated with it, which is not covered in this text. It is mentioned here only to let the reader know that it does exist.)

The third term that is used to characterize transmission lines and the matched conditions that may or may not occur on them is the *return loss*. This decibel value shows the difference in power levels between the input signal and the reflected signal (Figure 3.3). In Figure 3.3, a 0-dBm signal, P_i, is going into the transmission line. The power being reflected, P_r, is shown as -20 dBm. That indicates that the return loss for this particular system is 20 dB. As can be seen from the name, the return "loss" is a loss, so it is a negative value. The higher the negative value is, the better the match, that is, for a perfect match, the return loss ideally is ∞ dB. That, of course, is not possible, but if you get a return loss of 35 to 45 dB, you have an excellent match. Similarly, for an open circuit or a short circuit, all (or basically all) the energy is reflected back. That means the reflected-energy level is essentially the same as that being sent down the transmission line. Thus, the return loss for such cases is 0 dB. When the characteristics of a transmission line or a circuit are measured, the return loss usually is the parameter that is the easiest to measure, as will be shown in Chapter 4. That parameter then can be used to find the reflection

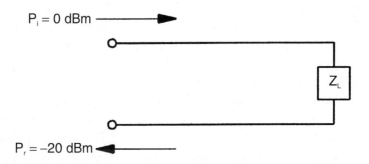

$P_i = 0$ dBm

Z_L

$P_r = -20$ dBm

Figure 3.3 An example of return loss.

coefficient, the VSWR, and finally the impedance of the load that is causing the reflections in the first place.

Now it is time to examine specific types of transmission lines. We will look at coaxial lines, stripline, microstrip, and coplanar waveguide. Each type of transmission line is described, along with definitions of associated terms, applications, and advantages and disadvantages.

3.1 Coaxial transmission lines

If you look up the word *coaxial* in a regular dictionary, you will find this definition: "having a common axis." That is descriptive, but it does not do much for our understanding of transmission lines in a coaxial configuration. If you look up *coaxial line* in an electronic dictionary, you will find the following: "a transmission line in which one conductor completely surrounds the other, the two being coaxial and separated by a continuous solid dielectric or by dielectric spacers." Figure 3.4 shows how those definitions apply to the transmission line we are discussing here. Figure 3.4 is an end view of a coaxial transmission line. You can see from the figure that there are three basic parameters to look at when describing a coaxial cable. The first is ε, the dielectric constant of the material used to separate the two conductors. Often, that material is Teflon®, which has an ε value of 2.1. Other times, it is polystyrene ($\varepsilon = 2.56$) or polyethylene ($\varepsilon = 2.26$). The dielectric is an important part of the coaxial line both mechanically and electrically. It is important mechanically because it provides the support for the center conductor and also the separation (or spacing) between the center conductor and the shield. It is important electrically because the type of dielectric determines the velocity of the electromagnetic wave traveling through the transmission line. The higher the dielectric constant, the slower the energy travels through the transmission line.

The second parameter in Figure 3.4 is the outside diameter, d, of the inner conductor of the transmission line, also referred to as the center conductor. The size of the conductor is an important part of the coaxial transmission line because it is used to determine many of the parameters for the line. The parameter d goes along with the third parameter D, which is the inside diameter of the outer conductor (usually the shield).

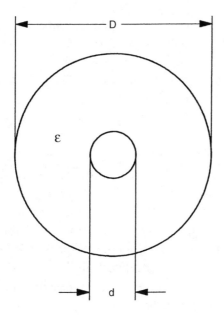

Figure 3.4 Coaxial cable.

Both d and D are important dimensions used in determining such parameters as impedance, capacitance, inductance, attenuation, and cutoff frequency. Each of these parameters relies on the dimensions of the center conductor diameter and the inside diameter of the shield for the transmission line.

As important as the two dimensions d and D are, by themselves they are not significant. They become critical when you form a term called the D/d ratio, which is used in virtually every electrical parameter associated with a coaxial transmission line. All the parameters we have mentioned (impedance, capacitance, inductance, attenuation, and cutoff frequency) rely heavily on the D/d ratio for their properties.

The importance of d (the outer diameter of the inner, or center, conductor) comes into play when you consider different transmission lines. For example, three particular transmission lines, RG-58, RG-59, and RG-62, look basically the same because their outside diameters are very close. This is one of those cases, however, where looks are deceiving. The RG-58 is a 50Ω line, RG-59 is a 75Ω line, and RG-62 is a 93Ω line. Since one of the major factors that determine the impedance of a coaxial transmission line is the D/d ratio of that transmission line, that means that since the inside diameters of the outer conductors basically are the same, there must be three different d values for the lines for the impedance to be different. In fact, that is the case. The RG-58 center conductor diameter is 0.031 in, RG-59 is 0.083 in, and RG-62 is 0.025 in. So make sure the coaxial cable you have is really the one you need for your application.

We will look at two types of coaxial transmission lines, or cables: flexible and semirigid. Each type has uses for specific applications. As we get into the specific types of transmission lines, it is important to remember that every transmission line is a lowpass filter, so each type of coaxial cable has an upper frequency limit. That number should be known so you do not use cable above that frequency. This is true for all transmission lines, whether they are coaxial, distributed lines (to be discussed later), or even waveguides, which are used for much higher frequencies. All transmission lines have a cutoff frequency.

3.1.1 Flexible coaxial transmission line

The dictionary defines *flexible* as "able to bend without breaking," an excellent definition for flexible coaxial cable. It is a cable that is able to bend in many different directions and that, if the minimum bend radius for the cable is not exceeded, will not break or affect the parameters of the cable in any way. This type of transmission line is one that can be used for applications in which the line must be bent around corners to make the necessary connections. It also is an excellent transmission line to be used in a laboratory environment, where many connect/disconnect operations are needed and where there is no standard way of connecting a cable to a piece of equipment. A laboratory probably is one of the toughest environments for any type of cable, and the flexible coaxial cable fills the bill nicely. The basic flexible transmission line, shown in Figure 3.5, consists of four basic sections: a center conductor (solid or stranded wire), a dielectric (usually Teflon), a braided outer conductor, and an outer covering. This construction is used for RG-58, RG-59, and RG-62 cables. Figure 3.6 shows a flexible coaxial cable. We have already seen that the center conductors are different to make the different impedances of each of the cables. The dielectric is basically the same, as

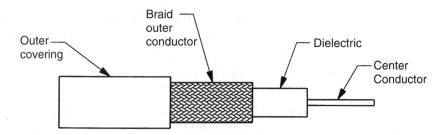

Figure 3.5 Basic flexible cable.

Figure 3.6 Coaxial connectors (Source: T. S. Laverghetta, *Modern Microwave Measurements and Techniques*, Norwood, MA: Artech House, 1988.)

are the braid diameters and the outer coating. This is where confusion can enter, because the visible portions of the transmission lines are so similar.

The center conductor can be either a solid wire or a series of wires in a stranded configuration. This is the "hot" lead of the transmission line, where the signal is carried. Also, it is the determining factor for the inductance and the resistance in the equivalent circuit. The center conductor usually is made of copper or a plated-copper configuration.

Because it is a metal and is carrying a current, an inductance is set up, and there also is a resistance simply because it is a metal and is carrying a current. The current is a varying one that is an electromagnetic current, that is, there is an electric and a magnetic component to it. Those two components are 90 degrees out of phase and form a complex current. Do not think of the current in an RF or microwave circuit as being something as simple as a dc current or an ac current coming from the outlets in your home. It is a complex current that cannot be measured with a meter. Thus, the center conductor of a flexible transmission line has an important job in carrying this complex current efficiently from one end of the transmission line to the other.

One of the tasks of the dielectric is its use as a spacer to keep the center conductor and the outer conductor apart, but that is probably the least important thing the dielectric does. It also determines many parameters of the transmission line itself. You will recall that this is an important parameter that was used to determine the capacitance and the conductance in the equivalent circuit. It is a vital part of determining those quantities. It also helps to determine the characteristic impedance of the cable, the capacitance per foot, the attenuation in the cable, the cutoff frequency, and the velocity of the RF and microwave energy propagating through the transmission line. So you can see that the dielectric material in a flexible transmission line does much more than simply keep two conductors separated. It is an important part of the entire system that we call the flexible transmission line.

The outer conductor serves two functions. It is a ground reference for the signal on the center conductor, as we can see from the equivalent circuit, and also is used as a shield. The shield keeps external signals out and internal signals in. The braid construction may be single, double, or triaxial (two braids separated by an insulator). The single-braid construction consists of bare, tinned, or silverplated copper wires. The double braid consists of two single braids with no insulation between them. The triaxial consists of two single braids with a layer of insulation between them. The type and the degree of shielding needed for your particular application determine which scheme is best for you. It may depend, in whole or in part, on the environment the transmission line is exposed to, the type and amplitude of signals that will be inside the transmission line, or cost factors. Whichever criteria are used, a variety of shielding types are available, and the proper one should not be difficult to find.

The outer coating provides protection for the cable. Such protection is mainly environmental. It plays no part in the electrical performance of the cable. It simply holds everything together and supplies the protection. The outer coating of the coaxial cable is what we were looking at when we said that the RG-58, RG-59, and RG-62 cables seemed almost identical. To see how close they actually are, consider Table 3.1, which shows all three cables, their impedances, their outer-coating dimensions, their outer-conductor dimensions, and the diameters of the center conductors.

Flexible cables are available in a variety of different types and usually have an RG designation (military designation for cables). The overall cable outside diameter can range from 0.078 in ($\frac{5}{64}$ in) to over 1 in. Many times, the size of the cable is dictated by the CW power it needs to handle. Other times, the size is dictated by such factors as installation area (does it need to bend around many corners?) or environmental conditions (both atmospheric and electrical). As in any choice, the application and its specifications dictate which cable to use. It should not come down to the idea that "we have used this cable for every application for the past five years, so we will use it for this one, too," probably the easiest way to get yourself in trouble. If you use a single cable for every application, you will find out very fast that no one cable will do every job. Each application should be looked at individually and judged by its particular requirements.

Table 3.1
Characteristics of Three Types of Coaxial Cable

Cable Type	Z (Ω)	Outer-Coating Diameter (in)	Outer-Conductor Diameter (in)	Inner-Conductor Diameter (in)
RG-58	50	0.195	0.150	0.031
RG-59	75	0.242	0.191	0.083
RG-62	93	0.242	0.191	0.025

Transmission lines are flexible and able to be bent around corners and into tight areas. However, even flexible cables have a minimum bend radius, that is, a radius beyond which bending the cables results in serious degradation of performance. Even though the cables are flexible, they cannot be tied in a knot and be expected to perform as well as when they are stretched out straight. Consult the cable data sheet to find out what the minimum bend radius is for each cable you use, even for test purposes in a lab (perhaps, we should say, *especially* for test purposes in a lab).

Flexible coaxial cables have many applications, ranging from finished RF cables for equipment shipped to a customer to the typical 50Ω cables hanging in every electronics lab. The performance of the final, assembled flexible transmission line depends not only on the flexible cable but also on the connectors placed on each end. There are many types of connectors that can be attached to a flexible cable. Four typical ones are shown in Figure 3.7. The *subminiature-A* (SMA) connector, shown in the upper left corner of the figure, is used for many microwave applications and comes in a variety of configurations, such as with two- and four-hole flanges for attachment to a chassis. The one shown in Figure 3.7 is the type used for cable connection.

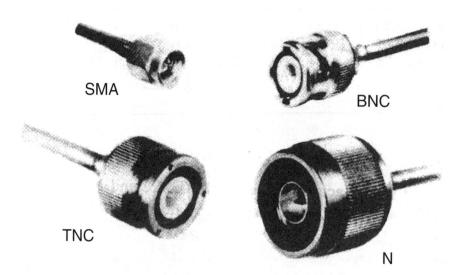

Figure 3.7 Coaxial connectors.

The connector shown in the upper right corner of Figure 3.7 is a BNC connector, the type seen most often on flexible cables in electronic laboratories. BNC connectors are the ones hanging on the wall in a college lab or on the bench in a company lab. They are good connectors for low-frequency applications. Higher frequency circuits and systems should not use BNC connectors because the ground connection with this type of connector is not the best when the frequency increases.

An improved version of BNC connector used for higher frequency applications is the connector in the lower left corner of Figure 3.7, the TNC connector. It is actually a BNC connector with threads, which makes a much better ground connection at higher frequencies. The connector shown in the lower right corner of Figure 3.7 is a type N connector, which is a larger connector that finds a wide variety of applications over a large range of frequencies. It also has threads for ground connection and can be fitted to a number of flexible cable types.

A great deal of care must be taken in the choice of connector to go on the ends of a flexible cable. The wrong connector can completely cancel all the good things you looked for when you chose the cable. The *right* connectors will make your choice of cable look that much better.

3.1.2 Semirigid cable

The dictionary defines *rigid* as "not bending or being flexible, stiff." What we need to understand is what we mean by *semirigid*. We can envision something that is fairly solid but capable of being bent. When you first see semirigid cable, you would not think of bending it. It looks and feels rigid. It does, however, bend to a specific bend radius but only *once*. We emphasize the word once because a piece of semirigid cable that has been bent should not be straightened or rebent. If the cable is bent in the wrong place, leave it alone and go from there. The best thing to do is to put the cable aside (and hope you can use it elsewhere) and bend a new piece to fit the application. If you attempt to straighten out a piece of semirigid cable, more than likely you will change the physical properties of the cable and thus change the electrical properties. The largest change that can occur is that the center conductor will have a bump in it rather than the nice smooth bend that proper bending produces. Also, the impedance will be different at the bump and cause reflections and a VSWR at that point that degrades the overall cables performance. Semirigid cable, like flexible cable, has a minimum bend radius. Consult the data sheet from the manufacturer to find out what that radius is. It actually is much easier to conform to the minimum bend radius of semirigid cable than to that of flexible cable because test fixtures are available for bending semirigid cable that already have the minimum radius on them. Nevertheless, to be safe, you still should consult the manufacturer's data sheet.

Semirigid cable is similar in many respects to flexible cable and also very different. To clarify that statement and to understand the idea behind semirigid cable, refer to Figure 3.8. Semirigid cable consists of a solid center conductor, a solid dielectric, and a solid outer conductor. Notice that every part of the semirigid cable is solid. That is what gives it its rigid appearance and characteristics.

By looking at Figures 3.8 and 3.5, you can see the similarities between semirigid and flexible cables. There is a coaxial construction for both transmission lines: a center conductor is surrounded by a dielectric with an outer conductor used for ground over the entire line. All these properties can be found in both the semirigid and the flexible versions of this type transmission line. Figure 3.9 shows typical semirigid cable.

The main difference between semirigid cable and flexible cable is the solid outer conductor for the semirigid cable. The solid conductor improves the shielding and gives the cable much more rigidity, which is exactly what it is intended to do. The solid outer conductor might be

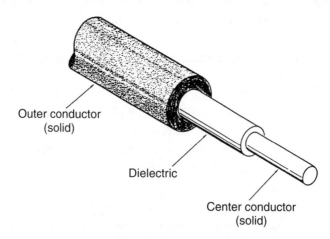

Outer conductor
(solid)

Dielectric

Center conductor
(solid)

Figure 3.8 Semirigid cable.

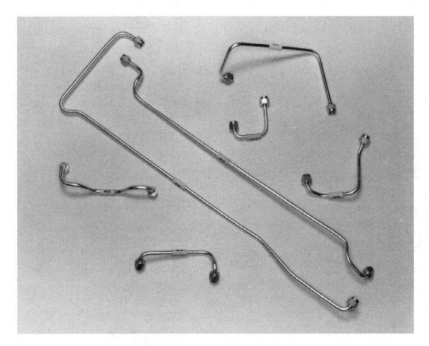

Figure 3.9 Semirigid cables.

simply a copper sheath over the entire cable, or it might be plated for specific applications. Usually the cable is gold plated to reduce oxidation of the copper over time, adding to the performance of the cable and also to the price. Other cables may be tin-lead plated for the same purpose. Some copper semirigid cables have been used for lab applications for some time. They are much darker in color because they have oxidized. They also look a little strange since they have been bent and rebent many times to conform to a specific test setup. (Because these semirigid cables have been bent more than once, their performance is far from optimum.)

Flexible cable is designated by an RG number (RG-58, RG-59, etc.), whereas semirigid is designated by a number only, that number being the outside diameter of the cable. Available sizes (in inches) of semirigid cable are 0.020, 0.034, 0.047, 0.056, 0.070, 0.085, 0.141, 0.215, 0.250, and 0.325. The three sizes most widely used are 0.085, 0.141, and 0.250. The large majority of semirigid cables in equipment are 0.141 cables, the beautiful gold paths that snake their way through equipment, connecting all the high tech designs to form a smooth-running system.

The other two common types of semirigid cable, 0.085 and 0.250, have specific areas of use for which the 0.141 may not be suitable. The 0.085 cable is used when there is very little room in which to maneuver, so the bends in the cable need to be very small. This small-diameter cable fits the bill nicely but exhibits higher attenuation than its 0.141 counterpart. The power-handling capability of the 0.085 cable also is less than that of the 0.141, which is understandable, since it is much smaller and there is less area to dissipate power in the 0.085 cable.

The 0.250 semirigid cable is used primarily for higher power applications, because it has much more area to dissipate power than either 0.085 or 0.141 cable. For a comparison, consider the power-handling capability of the three sizes of cable at 1.0 GHz. The 0.085 cable is rated at 222W, the 0.141 cable is rated at 600W, and the 0.250 cable is rated at 1200W (each of these is CW power). It can be seen that the 0.250 cable is far and above the other two cables. The 0.250 cable also exhibits lower attenuation than either of the other two models of semirigid cables. If we once again take 1.0 GHz, we can see that the 0.085 cable has 0.187 dB/ft of attenuation, the 0.141 has .116 dB/ft, and the .250 cable has 0.073 dB/ft. So that is another advantage of 0.250 semirigid cable. The big drawback of 0.250 cable is its size. It takes a lot of room to make a bend with a piece of basically rigid cable with a 0.25-in diameter. Thus, 0.250 cable is used basically for straight runs in high-power applications.

Semirigid cable has many superior electrical properties over flexible cable in many areas. In all cases, it can handle more power and exhibit less loss than a flexible cable of comparable size. For example, comparing a piece of RG-58 flexible cable and a piece of 0.141 semirigid cable at five frequencies, we find the data presented in Table 3.2. We chose these two cable types because their outside diameters are basically the same (the outer conductor of the RG-58 is 0.150 in and the .141 cable is, of course, 0.141 in).

Table 3.2 shows some very interesting numbers. If we look at the attenuation numbers at 3 GHz, for example, we find that the 0.141 cable has 0.215 dB/ft, while the RG-58 has 0.41 dB/ft, basically twice that of the 0.141. Similarly, if we look at the power-handling capability of the two cables at 3 GHz, we see that the 0.141 cable can handle 310W, while the RG-58 cable can handle only 22W. The 0.141 cable will handle approximately 14 times more power. We can also see that the RG-58

Table 3.2
Comparison of Semirigid and Flexible Cables

Cable Type	Attenuation (dB/100 ft) Frequency (GHz)					Power (W) Frequency (GHz)				
	0.4	1.0	3.0	5.0	10.0	0.4	1.0	3.0	5.0	10.0
0.141	7.2	11.6	21.5	18.5	44.5	1,000	600	310	230	160
RG-58	11	20	41	—	—	75	44	22	15	—

cable is not specified at all for attenuation at 5 or 10 GHz and is not specified for power-handling capability at 10 GHz. There really is nothing wrong with not being specified at certain frequencies. Recall that all transmission lines have an upper frequency limit since they all act like lowpass filters. That is clearly indicated in Table 3.2 for the RG-58 cable.

The numbers in Table 3.2 might lead you to ask, "Why ever use flexible cables?" The answer lies in a variety of areas other than the explanation of high-frequency limitations. First, semirigid cables cost considerably more than flexible cables, an important consideration, especially in the commercial market. Second, for testing applications, semirigid cables are not very practical. Most tests require many connect/disconnect operations, which can put strain on the cables. Also, as the tests change in a lab, there are different positions for the cables that are formed much better with flexible cables. Third, in some finished products, the cables must meander through the chassis to various locations. Semirigid cable would not fit those applications in many cases.

3.2 Strip transmission line (stripline)

The terms *strip transmission line*, *stripline*, *tri-plate*, and *sandwich line* all refer to the same type of transmission line. The term most commonly used is stripline. Stripline is different from, yet also similar to, flexible and semirigid cables. As a matter of fact, stripline actually evolved from coaxial transmission line and is very similar to it.

The stripline shown in Figure 3.10 appears to be similar to a coaxial cable; it looks like a coaxial cable that has been run over and flattened.

To realize how a coaxial structure can result in a stripline structure, refer to Figure 3.11. The coaxial structure shown in Figure 3.11(a) is a typical center conductor surrounded by a dielectric with a shield completely around the entire structure. There is nothing new about this. The new portion is shown in Figure 3.11(b), which is a side view of a stripline structure.

If we take the coaxial structure in Figure 3.11(a) and apply pressure at the top and bottom, as shown, the circular structure starts to deform and go into an oblong shape rather than a concentric circular form. The lines in the coaxial diagram are the electric field that goes from the center conductor to the outside shields of the structure. If we keep applying pressure to the coaxial device, eventually the two ends split and break. We then have a top ground (outside shield), a bottom ground, and a rectangular center conductor instead of the original circular structure shown in Figure 3.11(a). This is the same structure that was shown in Figure 3.10, the stripline structure. It evolved from the circular coaxial device and still has all the original sections (center conductor, dielectric, outside shield, and electric fields) but now is in a form that will operate at much higher frequencies and be more efficient for RF and microwave applications.

In both Figures 3.10 and 3.11, it can be seen that stripline actually is two circuit boards sandwiched together (hence the term *sandwich line*). There is a ground (copper) on the top and a ground on the bottom, with the actual circuit in between. The circuit is what is termed the center conductor in Figure 3.11 and is the straight line going through the center of the device in Figure 3.10.

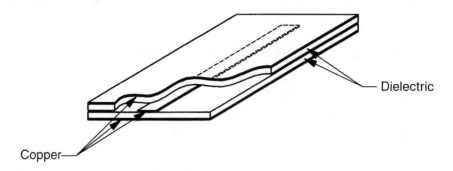

Figure 3.10 Cutaway view of stripline.

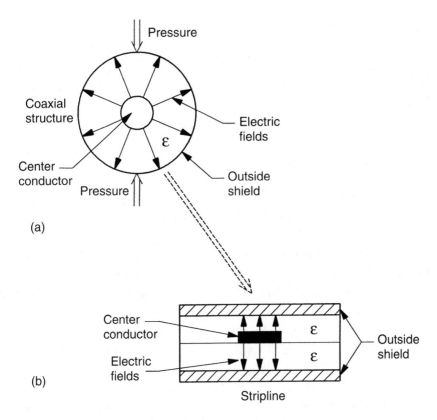

Figure 3.11 The evolution of stripline.

In the coaxial structure shown in Figure 3.11, the center conductor is surrounded by a dielectric material. In the stripline structure, each circuit board in the structure has a specific dielectric constant, which is the same for both boards, that is, the center conductor is surrounded by the same dielectric, top and bottom. In its simplest form, a piece of stripline would have the lower circuit board with the required circuit etched on one side and a complete ground plane (covering the entire circuit board) on the other. The top board would have a similar ground plane on one side and nothing on the other.

Before we go any further, it is necessary to clarify the term *ground plane*. In conventional low-frequency, dc, and digital circuits, a ground is a point on a chassis where all sorts of wires are run to ensure that everything is tied to the same point and has an adequate ground connection. In RF and microwave circuits, the waves being sent down the transmission lines are complex waves. To ensure that all those complex waves have a good ground reference, it is necessary to have an entire plane or large area of ground rather than a single point. That is why circuit boards have all their copper on the reverse side with the circuit etched on the top side. Also, a signal has all the same properties every half-wavelength and has the exact opposite characteristics every quarter-wavelength. If we place ground points on an RF or microwave circuit board, there is a good chance they would be separated by a quarter-wavelength at some frequency and our ground connection would be an open circuit instead of a short circuit. Thus, the ground plane is used for high-frequency applications.

Stripline is a uniform construction. Figure 3.10 depicts that uniformity, with the circuit being completely surrounded by the same dielectric on all sides. Because of that uniformity, stripline has a natural shielding effect on the circuit because it is completely enclosed, with metal on the top and the bottom. Actually, the stripline package also has metal on the ends and sides, resulting in a circuit being placed into a complete metal box and the excellent shielding of the circuit.

An important parameter in stripline is the *ground-plane spacing* (GPS), which is shown in Figure 3.12. GPS is, as the name implies, the spacing between the ground planes, or copper on the circuit boards. It is important to emphasize that the GPS does not take into account the thickness of the copper on the boards, only the thickness of the two pieces of dielectric.

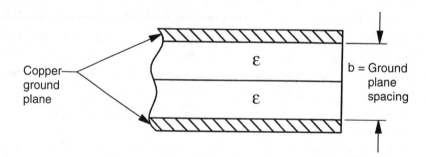

Figure 3.12 Ground plane spacing.

GPS, designated as *b*, is used to calculate the width of the transmission lines that are etched on the circuit board, *w*, and the spacing between two transmission lines, *s*. The relationships of *w*, *s*, and *b* are shown in Figure 3.13. The circuit board dielectrics must maintain a close tolerance on the overall thickness to ensure that the widths and the spacings are all as accurate as possible. (The tolerances on RF and microwave materials are covered in more detail in Chapter 6.) To understand how important it is to maintain the ground-plane thickness within a very close tolerance, consider the fact that to calculate the width of a strip transmission line you must calculate the *w/b* ratio. Similarly, to determine the spacing between two transmission lines, the *s/b* ratio is calculated. It is easy to see that if the *b* dimension in the denominator of the ratios varies widely it will be difficult to maintain a constant-width transmission line and, consequently, a constant impedance for the lines. Also, it can be seen that if the *b* dimension varies at all, the spacing between transmission lines will vary and the coupling between the lines will not be what was expected. That will cause a circuit to operate improperly.

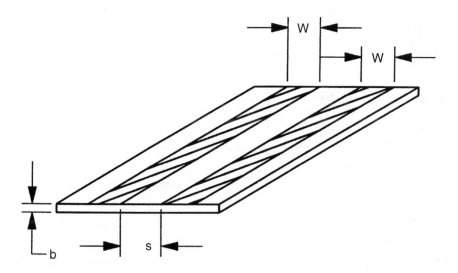

Figure 3.13 Relationships of GPS (b), transmission line width (w), and spacing between lines (s).

Stripline has become one of the selected methods of constructing transmission lines for many RF and microwave circuits. The only drawback to using stripline for some applications is also touted as an advantage. The advantage/disadvantage is that stripline is all closed up, that is, the circuit is completely enclosed by a dielectric material. Even though that is good for shielding a circuit, it makes it difficult to get at the circuit if a problem arises. To make matters worse, stripline packages many times have the boards laminated together and the packages sealed with epoxy. That brings about the rule of thumb that if an active device (transistor or diode) is used for an RF or microwave circuit, it generally is not built into stripline. If, of course, the module you are building is a throwaway module, there is no problem. If, however, you would like to be able to repair any problems that occur, stripline technology probably is not the best choice.

For a large majority of other applications, stripline is an excellent choice as a high-frequency transmission line. Whenever there is a requirement or desire for natural shielding of a circuit, strip transmission line should be used. Also, stripline is used for many passive components, that is, components that do not require a dc voltage to operate. (Passive components are covered in Chapter 4.) Figure 3.14 shows stripline circuits.

3.3 Microstrip

Microstrip transmission line does away with the problem of inaccessibility that stripline poses. Microstrip transmission line, shown in Figure 3.15, is similar to stripline transmission line, except that there is no top on the transmission line. There is nothing but air on top of the circuitry and a dielectric material underneath. In Figure 3.15, the width of the transmission line is designated by w, the thickness of the copper of the circuit trace by t, and the thickness of the dielectric by b. It can be seen that b is similar to that for stripline except that it is only one thickness of material, whereas the b for stripline is two thicknesses of material.

The thickness of the material for microstrip is just as important as it was for stripline. It is also used to calculate widths and spacings for transmission lines that are etched on the material used for microstrip. The

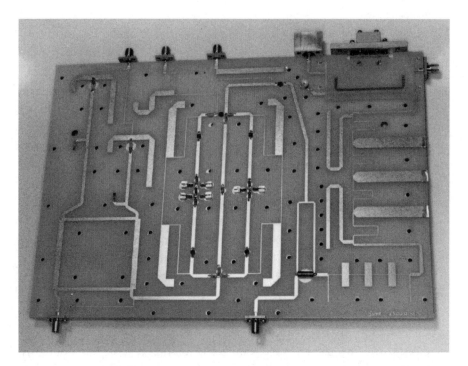

Figure 3.14 Stripline.

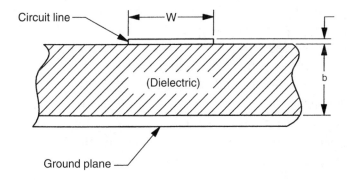

Figure 3.15 Microstrip transmission line.

same relationships apply to microstrip as they did for stripline in that w/b and s/b ratios are calculated design parameters. The main difference is that, as mentioned above, the *b* dimension for microstrip is half that of stripline. The stability of the thickness is, however, just as important for microstrip as it is for stripline.

Another parameter that is different in microstrip than it is in stripline is the dielectric constant. In stripline, the material (dielectric) surrounding the circuit line is all the same. Stripline is a well-behaved form of transmission line because it is arranged in such an orderly manner. For microstrip, it is a much different situation. If you look at Figure 3.15, you will see that, just as in stripline, the circuit has a very nice dielectric material under it. However, contrary to stripline, microstrip has only air above the circuit, not another dielectric, as in stripline. The boundary between the dielectric material and the air has been the subject of numerous research projects and papers over the years. It has been analyzed to the limit over and over again. The interface, however, can be simplified if you consider that you have two dielectric constants coming together, and there is a resultant dielectric constant to be used.

To compensate for the difference between the top and the bottom of the circuit line, we use an *effective dielectric constant*, ε_{eff}, which is actually the resultant dielectric constant. It is the result of a calculation that takes into account the dielectric constant of the material, ε_r, and the dielectric constant of air, $\varepsilon_o = 1$. The value of the dielectric constant is used for all calculations where a dielectric constant term is used. Note that every time the impedance of a microstrip line changes, the effective dielectric constant to be used changes, too. That is true because the width of a microstrip changes with a change in impedance. The lower the impedance, the wider the line, and vice versa. To calculate a filling factor for microstrip, which is a compensating factor for the difference in dielectric constant, the w/b ratio is used. Thus, a change in impedance is a change in filling factor and, consequently, a change in effective dielectric constant. The relationship between impedance and effective dielectric constant is a point that sometimes escapes people who do not note impedance changes very closely. This happens many times in the design of a multisection power divider or directional coupler. Such devices require different impedance lines that are all one quarter-wavelength long. The only problem that some people run into is that they do not calculate a new wavelength using each effective dielectric constant each time the impedance changes. That results in components with marginal performance. (Power dividers and directional couplers are covered in detail in Chapter 4.)

Microstrip transmission lines probably are the most common types of transmission lines that are visible in many of the RF and microwave circuits being fabricated today. The wireless markets use many microstrip circuits so components can be placed on top of the circuit board and attached easily. Manufacturers that use *surface mount technology* (SMT) components use microstrip because of the ease of construction. You can see how much easier it would be to attach components to a circuit board that has all the transmission lines on top of the board and very visible. Figure 3.16 shows microstrip circuits.

3.4 Coplanar waveguide

Usually when we think of waveguide, we think of frequencies well up into the gigahertz range. We also think of those rectangular pieces of hardware that have flanges on them that need to be screwed together and that exhibit very low losses. Sometimes it is difficult to visualize the devices because there is no real center conductor for energy to propagate down. Actually the devices are the ultimate in waveguide because they

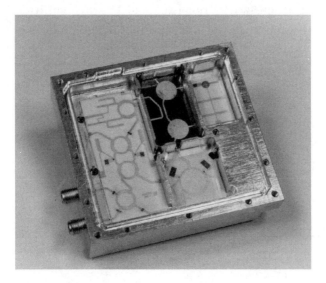

Figure 3.16 Microstrip assembly.

"guide" the wave down the structure. That is why no physical center conductor can be seen.

All those images are accurate and are what waveguide is. However, there is a much different type of waveguide: the coplanar waveguide, which is a modification of the microstrip circuit.

A representation of coplanar waveguide is shown in Figure 3.17. At first glance, it resembles microstrip construction. It has a single circuit board, just like microstrip; it has the circuit traces on the top of the board, just like microstrip; and it has air over the top of the circuit board, just like microstrip. When you look at it a little closer, however, you see some very distinct differences. In microstrip construction, there is a circuit trace on top of the board material of a certain width and thickness. There is also a complete ground plane on the reverse side of the board. In a coplanar waveguide, there is still a circuit trace on the top of the board that is a certain width and thickness, but there are also ground planes on both sides of the circuit trace and, as can be seen in Figure 3.17, there is no ground plane on the bottom of the circuit board. The ground plane on both sides of the circuit trace is where the transmission line structure gets its name. A conductor surrounded by ground "guides" the electromagnetic wave down the transmission line. It is also possible to construct coplanar waveguide as it is shown in Figure 3.17 but with a complete ground plane on the reverse side. This ground-backed coplanar waveguide is shown in a comparison drawing in Figure 3.18 along with conventional coplanar waveguide. Notice that the drawing in Figure 3.18(b) is exactly the same as that in Figure 3.18(a) except that

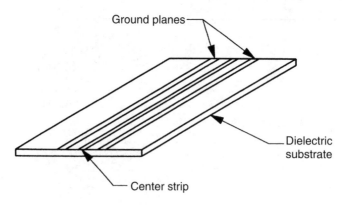

Figure 3.17 Coplanar waveguide.

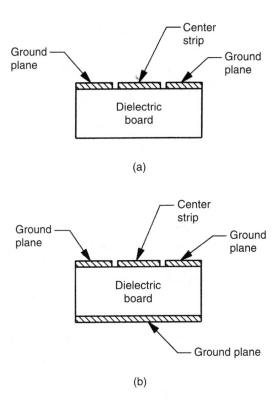

Figure 3.18 (a) Coplanar waveguide and (b) ground-backed coplanar waveguide.

the circuit in Figure 3.18(b) has a complete ground plane on the bottom of the circuit board.

One of the properties of coplanar waveguide that make it acceptable for RF and microwave applications is the fact that both series and shunt elements can be attached to the transmission line with relative ease. In the case of microstrip or stripline, the fabrication of such elements as series capacitors is a difficult task. With the coplanar waveguide, the ground plane is right next to the center strip carrying the signal, and connections to it are easy. You can see from the structure that it would be a rather simple task to connect a capacitor or a resistor from the center conductor to ground or to place similar components in series in the line. The large advantage of coplanar waveguide is that everything is accessible to the designer or fabricator. To make some basic connections, it is a matter of simply finding the most appropriate method of attachment for an RF or microwave component to do the particular task the circuit needs to accomplish and then making the connection. Coplanar waveguide offers some advantages that other transmission lines do not and is finding more and more applications in RF and microwave applications.

3.5 Summary

This chapter concentrated on the different types of transmission lines used in RF and microwave circuits. First, it covered the basics of transmission lines and defined such parameters as VSWR, reflection coefficient, and return loss. Next, coaxial transmission lines were discussed. Both flexible and semirigid cables were covered and comparisons made between the two.

The chapter then presented the types of transmission lines that use a dielectric circuit board as a base for their operations. The first transmission line described was stripline (also called strip transmission line, tri-plate, and sandwich line). This transmission line is a completely enclosed package that provides excellent performance and built-in shielding.

The next topic was the microstrip transmission line configuration. An area of interest in the microstrip discussions was that of effective dielectric constant, which results from the dielectric-to-air interface.

Discussion of coplanar waveguide and ground-backed coplanar waveguide rounded out the many different types of transmission lines that can be used for RF and microwave circuits.

The transmission lines covered in this chapter illustrate the need for special transmission lines for RF and microwave use. With these high frequencies, it no longer is possible to think of an interconnection between two circuits as simply a piece of wire that is wrapped or soldered to make the connection. There is a need for a low-loss type of connection to be used in this range, and that low-loss connection is the transmission line.

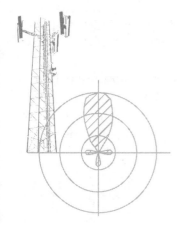

4

Microwave Components

WHENEVER YOU BEGIN to put together a jigsaw puzzle, you first lay out all the pieces. Usually you get all the edge pieces together and then maybe group colors or specific scenes. What you are doing is organizing and learning all the individual components that make up the puzzle. If you understand the components, you will have a much easier task of putting the puzzle together.

Likewise, if you are putting together a swing set for your children, it is a good idea to understand what all the pieces do before you attempt to assemble the unit. Even though few people actually do that, it is still a good idea. Most times people follow the old saying, "When all else fails, read the instructions." That is not a good habit to get into, whether you are putting together a puzzle, building a swing set, or assembling an RF or microwave system.

If you know what each individual component in a system does and how it operates, it is much easier to understand the overall system and how it works. That is how this chapter approaches the topic of microwave components. It presents a variety of components, explains their operation, defines terms applicable to them, and shows some typical examples of each.

Components covered are directional couplers, quadrature hybrids, power dividers, detectors, mixers, attenuators, filters, circulators and isolators, and antennas. You will see that it is much easier to understand an RF or microwave system once you know the basic operations of the components that make up the system.

4.1 Directional couplers

To understand the operation of a directional coupler, you must understand the terms *directional* and *coupler*.

First, let us look at the term *coupler*. If two transmission lines are placed close together, energy will be "coupled" from one line to the other, as shown in Figure 4.1. In the Figure 4.1, two transmission lines (A and B) are spaced a certain distance apart (*S*). There is no direct connection between the two lines. If the spacing is small enough, some of the energy applied to transmission line A will be seen on transmission line B. The amount of energy on transmission line B depends on how close the lines are to each another. The closer they are, the more energy that is *coupled* to the second line, transmission line B. The farther away they are, the less energy that is coupled. Thus, we can say that a coupler is a device that consists of two transmission lines with no direct connection, placed very close to each other so that a portion of the energy in one line is present in the second line. For most applications, that is not really what you want in an RF or microwave circuit. Usually you make every effort to keep

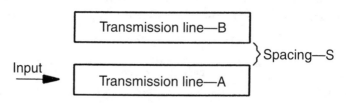

Figure 4.1 Coupled transmission lines.

transmission lines away from one another so that coupling does not occur. However, when you want to sample energy to check the power level or frequency of a specific signal, a coupler is an ideal component for the task.

The term *directional* means, basically, that energy is passed in one direction only. Any energy from a reverse direction will not pass or allow the component to operate properly. With the idea of what a coupler is, let us make that coupler directional. Figure 4.2 is a representation of a directional coupler. You can see that we still have two transmission lines (A and B) separated by a spacing (S). This is exactly what was shown in the coupler in Figure 4.1. The difference between Figures 4.1 and 4.2 is the term *L*, which is a specific length and makes the coupler directional over a certain band of frequencies. *L* is a quarter-wavelength at the frequency of operation. Generally, it is thought that the frequency of operation to be used is either a single frequency, if that is what you are using, or the center frequency of a band if you are operating over a band of frequencies. In reality, it is a good idea to make the quarter-wavelength value for a frequency that is just a bit higher than the band center. If, for example, you are operating at 800 to 900 MHz, you probably should calculate the quarter-wavelength at a frequency of around 860 MHz rather than the band center, 850 MHz. That ensures that the circuit will stand a better chance of operating at the high end of the band. If you had the quarter-wavelength at the band center, more than likely the response at the high end will be at a lower amplitude than if you used a slightly higher frequency.

To understand why this length of transmission line makes the coupler directional, recall how a signal repeats itself every half-wavelength and is exactly opposite every quarter-wavelength. A high impedance at one end

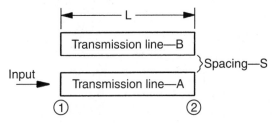

Figure 4.2 A directional coupler.

of a quarter-wave transmission line results in a low impedance at the other end. Relating to Figure 4.2, we can see that if the energy enters the coupler at the left of the device (point 1), it very easily will couple energy. As the energy travels down the line, it encounters higher and higher impedances and cannot couple any significant amount to transmission line B. When it gets to the end of the line, it has traveled exactly one quarter-wavelength and the initial low impedance is now very high so that virtually no energy is coupled. This is the forward operation of the coupler, which transfers energy to the second transmission line at point 1 easily.

If we put the energy in at the right side of the device (point 2), it will couple easily up to transmission line B because of a low impedance. Once again, as the energy moves down the line (from right to left), it encounters a higher and higher impedance until it gets to the left side (point 1) and virtually has an open circuit staring it in the face. Basically, no energy gets to transmission line B at point 1.

What has just been explained is the concept of a directional device. If we want to couple energy out at point 1 on transmission line B, it will happen readily. Any energy coming from point 2 will not couple out at point 1. The same would be true if we wanted to apply energy at point 2 and couple it out at that point. That would be a fairly easy task to perform. As in the preceding case, any energy coming from point 1 toward point 2 would not be coupled out at point 2. This is truly a directional coupler. It couples energy without any direct connection in one direction only. Figure 4.3 shows a directional coupler.

To further understand a directional coupler, we will define certain terms used to describe such a device. To aid in these definitions, refer to Figure 4.4, which is a schematic representation of a directional coupler. This is a four-port coupler. (A three-port coupler can be drawn by placing a termination, usually 50Ω, at port 4 and using the other three ports only.) The input is placed at port 1.

The energy path from ports 1 to 2 is called the *insertion loss*, which is usually in the range of 0.3 to 0.5 dB for higher coupling values. The insertion loss is determined by how far the coupled line is from the main input line. (If the spacing is very small, that could be as large as 2.0 to 2.5 dB.) Insertion loss many times is classified as a straight-through loss because the only thing between the input and the output is a single

Figure 4.3 Directional-coupler.

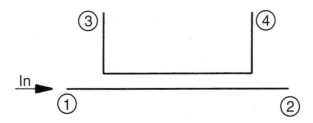

Figure 4.4 Schematic representation of a directional coupler.

low-loss transmission line. Sometimes you will see a width of transmission line that is different from the width of the input and output lines. That is usually the case with lower values of coupling. It is necessary to match the impedance of the device in the coupling area. Most couplers with coupling above about 10 dB have basically the same width of lines, and the insertion loss is the loss through that single transmission line.

The energy path from ports 1 to 3 is the coupling. This is the amount of input signal that appears at port 3 (in decibels) compared to the total energy that is at the input. The smaller the spacing is between the two transmission lines, the lower the value of decibels is read at port 3. The farther away they are, the higher the decibel value. Typical values of coupling are 10, 20, and 30 dB. There is usually a limit, determined by fabrication schemes, as to how close the lines can be to one another. If you are etching stripline or microstrip lines, there is a limitation in the etching of very narrow gaps between transmission lines. The value of coupling is typically limited to about 5 dB. This type of side-coupled circuit cannot usually be fabricated any closer with any sort of accuracy. The same is true of a coaxial system. Normally, a 6-dB coupler is as small as you can practically get with coaxial construction. (It should be pointed out here that when you have these low values of coupling in a device, you have approximately a 2-dB insertion loss compared to the 0.5-dB loss for higher coupled devices.)

The energy path from ports 1 to 4 is called the *isolation*. The isolation tells you how good a directional coupler you have. Ideally, there should be no energy at all at port 4 when the input is at port 1. In reality, however, there is a value that can be read. The isolation also is measured in decibels and can range from 20 to 50 dB, depending on the design of the coupler.

One term for directional couplers that many times is confused with isolation is *directivity*. Directivity is a measure of how good the directional coupler really is. It is not a measured quantity but is derived from the other measured parameters. It is the difference between the isolation and the coupling (all quantities in decibels). If, for example, we measure the isolation as 50 dB and the coupling as 15 dB, the directivity is 35 dB. You can see the problem that may arise if someone confuses isolation and directivity. In this example, there is a difference of 15 dB. If you asked for a coupler with a directivity of 50 dB when you meant an isolation of 50 dB, the coupler would have to have an isolation of 65 dB, with a 15-dB coupler.

A couple of other terms should be covered. The first is *frequency*, which, when we are talking about directional couplers, is the range of frequencies over which all the specifications are valid. If your application is within the frequency range of the directional coupler specifications, there will be a good match and your circuit should work well. If not, find another coupler.

The second term to look at is *coupling deviation*. No device is going to be perfectly flat across even a narrow band of frequencies, so there must be some allowances for any variations that occur. If you look at a directional coupler data sheet and find that it has a coupling of 15 dB (±0.75 dB), you can expect the coupling port to vary from 14.25 to 15.75 dB. If you realize this in the beginning, you will not be surprised when you get one coupling at 14.5 dB and another value at 15.5 dB.

4.1.1 Monitor circuits

Now the question remains, "Where are directional couplers used?" One application for a directional coupler is shown in Figure 4.5. A monitor circuit is one of the primary applications for a directional coupler. The coupler in Figure 4.5 is monitoring the output of circuit 1 or the input of circuit 2. The reason a directional coupler is used for applications like this is that it couples only in the forward direction (if there are reflections from circuit 2, they will not be seen at the coupled port because of the directional properties of the coupler) and has a very low loss through it (insertion loss). This last property is good because if you are going to put any device in a circuit for the express purpose of monitoring a certain point, you do not want it to interfere with the normal operation of that circuit. The device you use must be able to monitor the device effectively but still appear invisible to the circuit that is being monitored. This is what a directional coupler does very well. Recall that the straight-through port in the directional coupler is the insertion loss of the coupler and is a very low loss if the coupling factor is 10 dB or greater. Thus, the coupler exhibits a low loss between the circuits being monitored and basically appears as if it is not there at all, which is exactly what is needed for an effective monitoring device. Another property of the directional coupler that helps it appear transparent is the VSWR at the input. If the coupler is matched at the input, it exhibits a good VSWR for the driving circuit (circuit 1 in Figure 4.5). That also adds to the invisibility of the coupler in the overall system and makes things work smoothly.

The indication device in Figure 4.5 can be a power meter, a frequency counter, a spectrum analyzer, or a meter if a detector is used along with

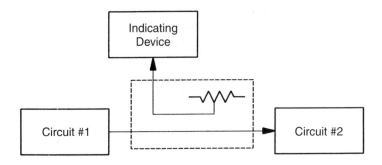

Figure 4.5 Directional coupler monitor.

the directional coupler. (Detectors are covered in detail in Section 4.4.) What we need here is some sort of device that will give us the indication we desire. If we need to know the power, a power meter will be there. If we need to know a proportional voltage for a corresponding signal level, a detector and a meter will be there. If we need to see the spectrum that is present at any particular point, the spectrum analyzer will be there. Be sure to remember, however, that the power levels you read at the indicating device are attenuated from the actual level of the circuit by the amount of coupling that the coupler exhibits. If, for example, we read a signal level of −22 dBm on a spectrum analyzer connected to the monitor in Figure 4.5, and the directional coupler is a 20-dB coupler, the actual level of the signal in the system is −2 dBm. This can be critical in the monitoring of actual power levels, so be sure to take it into account when you are actually determining what levels you have in a monitored system.

4.1.2 Leveling circuits

Another application of a directional coupler is one you may not see or realize is working in your system or test equipment. That application is a leveling circuit, which is shown in Figure 4.6. Many RF and microwave sources have such a circuit built in to them. Some older models of sources do not have a leveling circuit built in, and the output level falls off as the

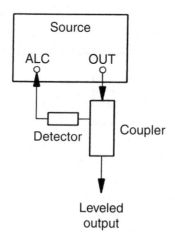

Figure 4.6 Directional coupler used as a leveling circuit.

output frequency gets higher and higher. The circuit arrangement used for either an internal or external leveling circuit shown in Figure 4.6 will provide the required leveling of the output signal.

If you have ever looked at the output of a sweep generator as it sweeps across a band of frequencies, you will realize how necessary a leveling circuit is. The output is usually at a pretty reasonable level when it starts out at the low end of the frequency band, but as it increases in frequency you can see the level drop off. Sometimes the drop is not that significant, but other times it is something that needs to be taken care of. Many times it depends on just how wide the band of frequencies being swept is. Generally, a leveling circuit is used to ensure that a flat response is present for testing circuits or for a flat response within a system.

To understand the leveling circuit, you must understand a basic feedback control system. To do that, let us look at the most basic of all feedback control systems, the human body. Say you are required to hold a weight out in front of you at eye level. You need to hold it level as long as you possibly can. If the weight starts to drop below the required level, your eyes (monitors and sensors) sense the change and send a signal to your brain (control center), telling it that the level is not right. The brain sends a signal to your muscles (control systems), telling them to lift the weight up. If the muscles overreact and lift the weight too high, the eyes once again sense this and tell the brain it is not right, the brain tells the muscles to lower the weigh, and the weight is once again at its proper level. Of course, a point may be reached where the brain tells the muscles to raise the weight and the muscles do not respond. Nothing the brain can do will correct for that "too tired to respond" condition. The basic idea of the feedback control system still holds true and can be applied to electronic systems and test equipment.

To understand how the system in Figure 4.6 works according to our basic example of the human body, we will describe each component in the setup. The signal coming from the source is monitored and sampled by the directional coupler as if it were in a regular monitoring application, as presented in Figure 4.5. The output from the coupled port of the coupler has a detector connected to it. The detector converts the signal level to the appropriate dc voltage. The voltage is then sent to an *automatic leveling circuit* (ALC), which is a voltage-controlled attenuator at the output of the source. If the output level drops, the coupler-detector combination senses that and sends a voltage to the attenuator, which removes attenuation and increases the output level. If the output level increases, the process is reversed, and attenuation is inserted in the output line. Leveling circuits have been used in many pieces of test equipment and in many systems for a long time. It is an effective and simple leveling system that simply takes advantage of basic feedback principles.

4.1.3 Power measurement

Another application of a coupler is as a means of measuring power. It is used when you have a higher power to measure, and the power meter, spectrum analyzer, or counter cannot handle that high a power level. The coupler acts to attenuate the power and make it safe to read on the meter or analyzer you have (e.g., you have a half-watt of power to read, but the power meter you have can be used only up to +10 dBm). The arrangement in Figure 4.7 can be used to allow that to happen. First, we must convert the half-watt to decibels referred to milliwatts, which is +27 dBm. If we use a 20-dB coupler in the setup in Figure 4.7, we will see +7 dBm at the meter or analyzer, which is below the maximum level of the indicating device to be used. Thus, you get your measurement and protect the power meter being used. It is a good idea to check the maximum input requirements to all indicating devices to be sure you do not blow out the front end of a spectrum analyzer or other device. That can get frustrating and expensive, both in cost of repair and down time for the instrument.

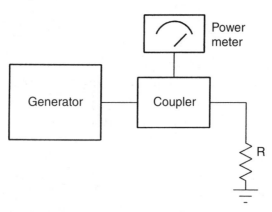

Figure 4.7 Directional coupler used to measure power.

4.1.4 Reflectometers

Our final application uses a special directional coupler, the *dual-directional coupler*. This application is a reflectometer and is shown in Figure 4.8. It can be seen that the dual-directional coupler is actually two single directional couplers placed back to back. The couplers may be physically joined with an adapter or actually come in a single package from a variety of manufacturers. What the dual-directional coupler does is provide two output ports with identical coupling (ports A and B) and an increased isolation between those ports (also an increase in directivity). The large isolation is an absolute necessity in a reflectometer setup. The dual-directional coupler also changes other parameters when they are put back to back. The insertion loss increases because there are now two couplers instead of just a single coupler. The insertion loss goes from approximately 0.25 dB for a single 20-dB coupler to 0.75 dB for the dual-directional coupler. Also, the amount of reflected power that can be handled by the coupler increases because of the isolation that takes place in the coupler itself. If, for example, the forward power level that can be handled by a single or dual coupler is 50W, the reflected power for the single coupler would be approximately 10W, while the reflected power that could be handled by the dual coupler would be 50W, because of the increased isolation in the dual-directional coupler.

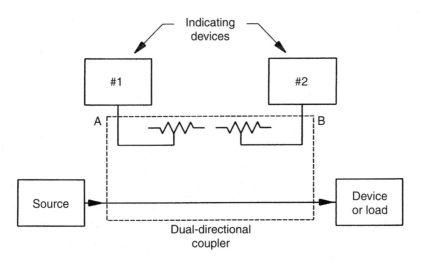

Figure 4.8 Dual-directional coupler used as a reflectometer.

A reflectometer measures the input power to a device or a load and the reflection from the load, compares them, and gives the return loss of the device or load in decibels. The input power is measured by indicating device 1 and the reflections by indicating device 2. This setup results in a characterization of a particular device or load with regard to its match compared to the system characteristic impedance. It does that by taking advantage of the increased isolation and directivity of the dual-directional coupler. The increased isolation means that when the forward power is measured at indicating device 1, that is the only power being read. There is no reflected power because of the outstanding isolation between the ports. Similarly, when you read the reflected power with indicating device 2, you are measuring only the reflections, not the forward power. With this arrangement, you can be sure you have an accurate reading and your measurement system is accurate and repeatable.

It can be seen that the directional coupler is a versatile component that finds use in many areas of RF and microwaves. It is valuable to be able to place a component in a circuit to monitor a specific location and not have to worry about a dc path for its operation. Also, the directional properties of the coupler help to separate signals within a system. The only area you have to watch is that you have a coupler that is in the proper frequency range for your application.

4.2 Quadrature hybrids

The quadrature hybrid is another component that is more easily understood if you define the individual terms of its name. Just as with the directional coupler, doing that adds clarity to the explanation. The first term to define is *quadrature*. If you look up this term in the dictionary, you will find that this definition: "any two objects that are at right angles to one another." If, for example, you have two lines at right angles (90 degrees), they are in quadrature; if you have two tables at right angles, they are in quadrature; if you have two people at right angles, they are in quadrature. What we have in quadrature in the quadrature hybrid are the two output signals. They are 90 degrees out of phase with one another in quadrature; that is, there are two outputs from a quadrature hybrid, and if you measure the phase relationship between them you will find a

90-degree difference. This is a valuable relationship to have, as we will see later on.

The term *hybrid* is actually the term *hybrid junction*. In that arrangement, transmission lines come together in such a manner that there is a high value of isolation between ports when it is used in a component. Figure 4.9 shows a quadrature hybrid and indicates the hybrid junction area. It can be seen that the hybrid junction is where the transmission lines overlap one another. When a quadrature hybrid is printed on a microwave circuit board, the hybrid junction is the only common point on both sides of the board material. Two 50Ω transmission lines are on one side of the board (one for an input and one for an output), and there also are two 50Ω lines on the other side of the board (also one input and one output). The only area that is on both sides of the board material is the junction area, or hybrid junction.

The hybrid junction provides excellent isolation between the input port and the port with the termination on it, and between output port 1 and output port 2. This is an important property for an RF and microwave component to exhibit because in many applications there is a need to keep signals apart (i.e., provide isolation between ports) to improve performance.

With the individual terms defined, now we can answer the question, "What, then, is a quadrature hybrid?" The quadrature hybrid is a special-

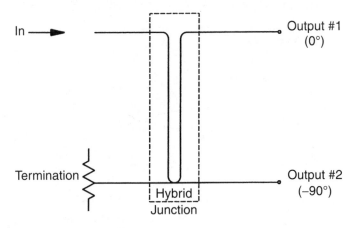

Figure 4.9 Quadrature hybrid.

ized coupler (note the coupling section at the hybrid junction where energy is coupled between two transmission lines) that takes an input signal and equally divides it between its two outputs. Ideally, the two outputs will be 3 dB below the input power signal level. The outputs are also 90 degrees out of phase, which allows for the construction of other components.

The quadrature hybrid has other features that can be put to use in many RF and microwave applications. If, for example, you apply equal signals to the two output ports, as they are shown in Figure 4.9, the two signals combine at the input of the device, with a small insertion loss taking place during the combination process. So, what we have is an accurate and efficient 3-dB coupler (a decrease in input power of one-half is 3 dB) that has some unique properties.

In Figure 4.9, the input signal is applied at the left. The other port on the input side is terminated in 50Ω. Output 1 is one-half the amplitude of the input signal (−3 dB) and is in phase with the input signal. Output 2 also is one-half the amplitude of the input signal and 90 degrees out of phase with it. Figure 4.10 shows a plot of amplitude versus frequency for the two output ports. It can be seen that output 1 has a plot that is bowed down over the frequency range of operation. Similarly, output 2 has a plot that is bowed up over the frequency range. Combining the two plots shows the operation of the entire quadrature hybrid. Points A and B are where the two output plots cross on the low-frequency end and the high-frequency end, respectively. Those two points show the frequency

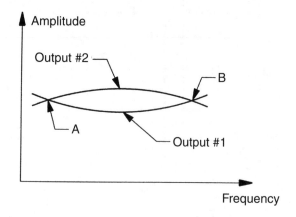

Figure 4.10   Plot of amplitude vs. frequency.

band of operation for the quadrature hybrid. If you take the frequency at point B and subtract the frequency at point A, you have the bandwidth of the component. Ideally, those two points would be at a level of −3 dB. In reality, however, that point is around 3.15 to 3.2 dB, which indicates an insertion loss for the quadrature hybrid. The insertion loss is the difference between the measured losses and the ideal losses at those points. In this case, the insertion loss is 0.15 or 0.2 dB.

There is a high degree of isolation between the two output ports and the two input ports of a quadrature hybrid (because of the hybrid construction). This excellent characteristic of the quadrature hybrid allows different signals to be applied at the input of the device without our worrying about them interacting with one another. It also ensures that the two outputs of the device will not interact either. That is a rare characteristic for an RF and microwave component to exhibit and is used to the maximum benefit in a variety of components in which the hybrid is an integral part. Figure 4.11 shows a quadrature hybrid.

To build some of those components, we can take two quadrature hybrids and put them back to back, as shown in Figure 4.12. A signal applied at the input basically appears at the same power level at the output, with the addition of a small insertion loss. The arrangement can be used to create individual components that will have good impedance

Figure 4.11 Quadrature hybrid.

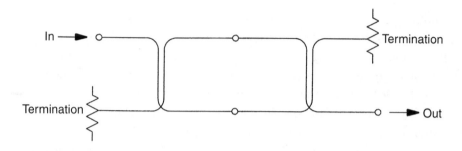

Figure 4.12 Combining hybids.

qualities. If, for example, a device (transistor, diode, etc.) is placed at the output port of the first quadrature hybrid and the device goes bad, the reflections from that failed device are sent back toward the input. Recall that the hybrid junction characteristics of the device are such that any signal sent back toward the input are deposited in the terminated port and do not affect the input signal at all. That is exactly what happens with the arrangement in Figure 4.12. This property results in a device that has stable impedance parameters and, thus, a constant VSWR at its input. Such an arrangement leads to this type of device being termed a *constant-impedance device.*

Two additional terms typically appear on data sheets for quadrature hybrids and should be defined and explained before we proceed any further: *amplitude balance* and *phase balance.* Amplitude balance is a comparison of the power levels at the two output ports. Ideally, there should be no difference in amplitude between the two ports. They both should be down 3 dB from the input level (or one-half the power of the input signal). In the real world, however, there will be some difference or variation. This figure should be as low as possible so each circuit that is driven by the quadrature hybrid receives the same amount of energy. A typical value for amplitude balance would be ±0.5 dB, which says that the amplitudes of the two outputs can range from 2.5 to 3.5 dB and still be within specifications.

The second term is *phase balance.* As stated earlier, the quadrature hybrid is designed to have a 90-degree phase separation between the output ports. Because of variations within the device, that phase varies with frequency. The phase balance specification is a measure of how well the phase difference tracks over the frequency range of operation. A typical value for phase balance is to have the output ports track within ±1.5 degrees. In other words, if the phase difference is 88.5 to 90.5 degrees, the hybrid is within specifications and will operate properly. The ideal phase balance, of course, would be 90 ±0 degrees. That may be possible over a very narrow band of frequencies, but it is not something that is practical to look for over a range such as an octave (e.g., 1 to 2 GHz). (An *octave* is a representation of a frequency band in which the high-frequency end of the band is twice the low-frequency end, e.g., 1 to 2 GHz, 2 to 4 GHz.) Thus, the phase balance should be as close to 0 degrees as possible.

4.2.1 Matched detectors

Now let us look at some applications of the quadrature hybrid. In Figure 4.13, a hybrid is used to produce a matched detector. This application uses matched diodes, which means the characteristics of each diode are the same. Matched diodes can be requested from a manufacturer and usually result in a higher cost, but if they are needed for excellent performance the cost is well worth it. It can be seen in Figure 4.13 that the two outputs from the quadrature hybrid feed the diodes. That means the signal to each diode is identical, within the specifications of the hybrid (amplitude balance). There will be some difference in amplitude, but it usually is insignificant to the operation of the detector.

The matched detector is designed to take an RF signal at the input and use the matched diodes to convert it, very efficiently, to a video or dc signal, as needed. The quadrature hybrid and matched diodes are what

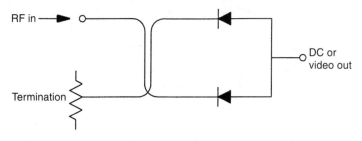

Figure 4.13 Matched detector.

make this circuit work as well as it does. The combination of basically equal output power levels from the hybrid, and the isolation between input ports and output ports make this a very good component to use when a detector is needed.

4.2.2 Combining amplifiers

Figure 4.14 is another quadrature hybrid application, one that uses back-to-back hybrids with some component in between. Recall that by doing this we have the same level at the output as we put in at the input, with a small insertion loss. If we put amplifiers between the two hybrids, the inputs to the amplifiers are fed with the equal amplitude output ports of the input quadrature hybrid. The outputs of the amplifiers similarly are applied to the two input ports of the output quadrature hybrid. Recall that an input signal is equally divided at the output ports. Also, if we apply two signals to the input, the hybrid combines the signals at the output port. This is what we are doing with the output signals of the two amplifiers. As a result, we have a small input power to the entire component and an output that is the result of the two amplifiers being used.

Although the operation presented here is a tremendous quality for this circuit to have, an even better one that makes this type of circuit useful for a wide variety of applications. That other quality keeps any mismatches produced by the amplifier stages from being seen at the input of the entire assembly. That is, if one of the amplifiers goes bad, there

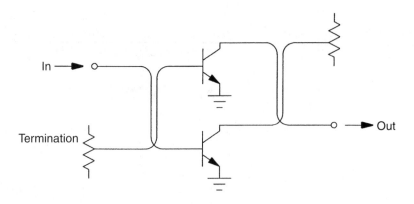

Figure 4.14 Combining amplifiers.

will be a very high impedance at the input of that amplifier. That causes a very high reflection that tries to get back to the input. Because of the hybrid junction characteristic of the component, the reflections all are dropped into the termination resistor at the other port at the input. Thus, the reflections are never seen at the input. This characteristic of a network with two quadrature hybrids placed back to back is a constant impedance device, which is a valuable characteristic used for many amplifier applications. It is ideal for obtaining higher power outputs as well as eliminating the worry of a bad amplifier causing an increased VSWR at the input of the device. You must remember, however, that if one of the amplifiers does go bad, there will be a reduced output level from the device just as if only one amplifier were being used. The big advantage is that the additional reflections caused by this amplifier are not seen at the input.

4.2.3 SPST switches

Another application of the quadrature hybrid is the *single-pole single-throw* (SPST) switch (Figure 4.15). This device also has a quadrature hybrid at the input and the output in the back-to-back arrangement. Diodes are placed in between that exhibit various values of resistance when biased. (These PIN diodes are covered in Chapter 5.) When the diodes are fully biased, they exhibit a low value of resistance, and the circuit appears to have nothing in between the two quadrature hybrids. For this condition, the energy goes from the input port, through the hybrid combination, and out the output 1 port with a very low loss. The loss will be in the order of tenths of a decibel, since there is really nothing to cause a loss other than the transmission lines of the hybrid and those connecting the hybrids.

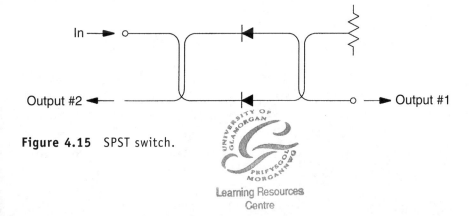

Figure 4.15 SPST switch.

When the bias is removed from the diodes, they exhibit a very high resistance, almost as if an open circuit has appeared between the two hybrids. This condition causes all the energy coming into the device to be reflected back toward the input. Because a hybrid junction is present, the energy does not get back to the input but is sent to the other port on the left in Figure 4.15, output 2. That results in a very low loss signal at that port. Thus, we have produced a very efficient switch. With one condition, we switch the energy to port 1, and in the other condition we switch it to port 2. This is a classic and very basic switch.

If we take the switch circuit and vary the diode bias continuously between full on and full off, monitor output 1, and place a termination at output 2 we would see various levels at that output port 1, depending on the value of bias. In one case, we produce a very small loss at the port and, as we continue to vary the bias, gradually pick up more and more loss at that port until we reach the point where there is no bias and a very high loss at the output. The type of component we have just produced is a continuously variable attenuator. The bias voltage for this type of component may be analog (continuously varying with respect to time) or a digital step (discrete voltage levels) to make a voltage-controlled attenuator. In the first case, we have what was just described. In the second case, we have a *step attenuator*, that is, each of the discrete voltages in a digital pulse produces a discrete value of attenuation within the device. Thus, the choice of a quadrature hybrid for this type of application produces valuable components that have very good qualities.

The quadrature hybrid is a valuable component for RF and microwave systems. Any place where isolation is needed between ports, where equal output powers are required, or where a 90-degree phase shift is necessary, the quadrature hybrid will be right at home and do an excellent job.

4.3 Power dividers

A quadrature hybrid takes an input signal and divides it into two equal outputs that are 90 degrees apart in phase, which is fine for some applications. But what if you do not want a phase shift between your output ports and you need to divide the input signal into three or four outputs? If that is the case, you can employ a simple 0-degree power divider that is not limited to just two outputs. *Wilkensen power dividers* provide a variety of output ports that are all in phase, within a reasonable tolerance.

Figure 4.16 shows a two-way, a three-way, and a four-way power divider. You can see from the figure that each of these dividers has an

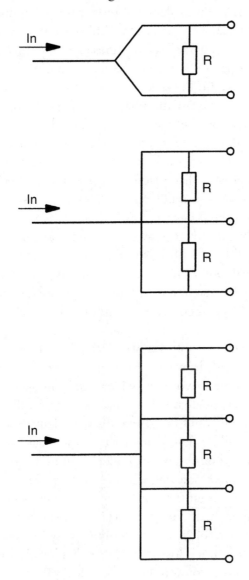

Figure 4.16   Power dividers.

input port on the left that comes to a point where the device splits off and goes in as many directions as is necessary for the individual component. It should be pointed out here that the input transmission line and the output lines will be 50Ω lines. The transmission lines for the division lines will be various values, depending on the number of outputs needed. For a two-way divider, the impedance of the transmission lines going to the 50Ω outputs is 70.7Ω. You can see the difference in widths of these lines as you look inside a circuit and see the lines that make up a power divider. Also, each divider has a resistor between output ports. The resistors are not there just to take up the space between the transmission lines, they are there to ensure good isolation between the ports. If the resistors are not there, the isolation will be very low. The division lines are each a quarter-wavelength long; if you do not terminate the lines, you will have the equivalent of close to a half-wavelength transmission line between any two output ports. That results in no isolation, because the ports basically are at the same point electrically. So, the resistors are important to the operation of the 0-degree power divider.

As with quadrature hybrids, when we divide an input signal into two outputs, there will be a 3-dB loss in the signal level. We also may have not only a two-way but a three-way, a four-way, and maybe more than that. To predict what the losses will be for each port, consult Table 4.1, which shows the ideal (theoretical) output levels associated with a given number of outputs.

The values given in Table 4.1 are ideal values if there is no insertion loss. In the real world, however, there will be an insertion loss, so the actual loss per port for a two-way divider will be in the range of 3.15 dB; for a three-way port, the loss will be approximately 5 dB; for a four-way port, the loss will be about 6.2 dB; and so on. Thus, the actual loss for each output port will be slightly higher than the ideal loss. That must be taken into account when you put together a complete system and look at all the losses and gains associated with the system. The difference we are talking about is only about 0.15 or 0.2 dB, but those small values have a habit of adding up to many decibels, which can cause problems later on in a system. It is possible that an amplifier may not operate properly because there is too much loss in a power divider used previously in a system. So, take into account not only the ideal value of loss but also the insertion loss that will be present. Figure 4.17 shows a power divider.

Table 4.1
Power Divider Outputs

Number of Output Ports	Loss (dB)
2	3.00
3	4.77
4	6.00
5	7.00
6	7.80
7	8.40
8	9.00
9	9.45
10	10.00

The terms *amplitude balance* and *phase balance* apply to in-phase power dividers as well as to quadrature hybrids. Amplitude balance tells how closely the output ports track one another. If, for example, one port of a three-way divider measures 4.98 dB, while the next port measures 5.00 dB, and the third port is 5.01 dB, the outputs track within 0.03 dB. That is an indication of the amplitude balance of the divider and is the same as would be specified for the quadrature hybrid. The values of amplitude balance you would see on a data sheet would be in the range of ±0.2 to ±0.5 dB (depending on the number of output ports).

The phase balance for a conventional power divider is somewhat different from that of a quadrature hybrid. Recall that the output phase for a quadrature hybrid is 90 degrees apart. For the 0-degree power divider, the output phases must be as close as possible to being in phase, or at 0 degrees. For the 0-degree power divider, the expected values of phase balance will be in the range ±1.5 to ±3 degrees. Applications of 0-degree power dividers range from use as signal dividers in test setups to mixer circuits, where signals must be fed in phase, to areas where multiple antennas must be fed in phase. Actually, any application where

Figure 4.17   Power divider.

there are multiple areas to be fed in phase call for the use of a 0-degree, in-phase power divider.

One application in which both quadrature hybrids and 0-degree power dividers are used is an image-rejection mixer, which is shown in Figure 4.18. An image-rejection mixer gets rid of any image frequencies generated by a mixer circuit. All we want out is the desired signal. As can be seen in Figure 4.18, there are two areas called balanced mixers, which are covered in detail in Section 4.5. For now, we will say that these devices have two inputs, RF and *local oscillator* (LO), and one output, *intermediate frequency* (IF). Each mixer is fed with RF and LO signals from the original circuits by means of some type of power divider. In the case of the RF input, it is the quadrature hybrid. That is because there needs to be a 180-degree phase shift between the two RF inputs to the balanced mixers. That is accomplished by using the natural 90 degrees from the hybrid and inserting a section of transmission line that will develop

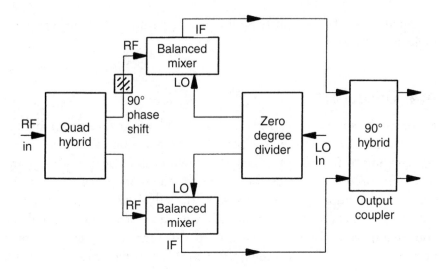

Figure 4.18 Image-rejection mixer.

the additional 90 degrees to make the total 180 degrees necessary for operation.

The LO port does not have the phase difference requirement, so the 0-degree power divider is used for this application. Once again, each of the balanced mixers receives the same amplitude signal, as well as the same phase.

Note in Figure 4.18 that at the output there is another 90-degree hybrid. That hybrid is the actual output of the image rejection mixer and is a quadrature hybrid designed to operate at the lower IF. This device allows the operator to choose either an upper-image or a lower-image frequency for an output. Once again, the quadrature hybrid pops up to do the job required.

The 0-degree power divider is a component that is used in many areas both by itself and with other RF and microwave components. It is a component that is fairly easy to fabricate until you get into higher-division applications. Then you need to do some serious thinking about how to lay out the device and still maintain the quarter-wavelengths necessary for proper operation. Also, the terminating resistors must be taken into account as to where they must be attached. Generally, however, there are many ways of accomplishing those feats, resulting in some very good dividers.

4.4 Detectors

Probably the simplest and most useful RF and microwave component is the detector. You can see how basic the construction of the detector is by referring to Figure 4.19. The first block in the diagram is an input-matching circuit that is designed to match the diode impedance to the characteristic impedance of the system (usually 50Ω). A matching network is necessary so the maximum power can reach the diode and the conversion from RF to dc can take place efficiently. That matching network has the task of matching the complex impedance (a value of real resistance and a value of some sort of reactance, usually capacitive) to the real value of 50Ω, which is probably the characteristic impedance of the system driving the detector. Many times the network is a series of transmission lines of various lengths and widths that provide the necessary matching for the detector. This network allows the RF energy to be transferred efficiently to the diode so the detection process can occur. Without the matching network, a great deal of power would be lost between the input and the diode circuitry, and the detector would be a very inefficient device that would find few applications.

The second block in Figure 4.19 is the diode itself (or a set of diodes) that actually performs the detection process. A detection process is one in which the full cycle of the RF input signal is clipped so that only the positive or negative portion of the signal is present at the output of the diode circuitry. This may be a half-wave rectification in which only

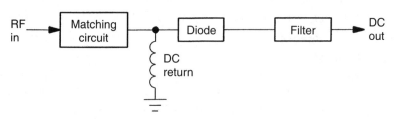

Figure 4.19 Detector.

positive "pulses" are present, with the opposite polarities being nonexistent, as shown in Figure 4.20(a). Another method is full-wave rectification, in which the positive or negative signals are reproduced for every cycle with the opposite polarity being transferred up (or down) to make a continuous signal output, with no gaps between the polarity being used, as shown in Figure 4.20(b).

The most important part of the detector is the diode. After all, the diode is what actually carries out the detection process and makes the entire component do its job. Another important part, however, is the dc return, which usually is a printed high-impedance transmission line that is a quarter-wavelength long and acts as both a dc return to ground and an RF blocking component. If we make a transmission line a quarter-wavelength long and attach one end to ground, we will have a very high impedance at the other end of the line (basically an open circuit). That is what the dc return does in RF and microwave detectors. It is a dc short for the current to return to ground, and it is also a high impedance for the RF that allows the blocking action. The dc return is critical, because the detector will not work unless there is a dc return for the diode. To

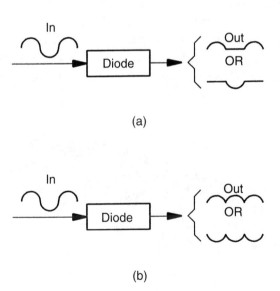

Figure 4.20 Detection processes (a) half-wave, (b) full-wave.

go back to basic diode theory, for the diode to perform and conduct current, there must be a complete circuit for the diode. That is where the dc return comes in; it completes the diode circuit.

The final block is a lowpass filter circuit that will result in the clean dc voltage at the detector output, and will remove any RF from the output if either a dc or video output is desired. This filter may be something as simple as a single capacitor that is attached to the output of the diode circuit and ground. This will make an adequate lowpass filter if the proper value is chosen. Generally you will see a conventional lowpass filter network put in this block. It will ensure that the output of the detector is the required dc, video, pulse, or any other output that you expect and not have any of the RF or microwave signal associated with it. Figure 4.21 shows a detector.

One term that is used to characterize detectors is *sensitivity*. Consider what you think of when you hear the word *sensitive*. One image that may come to mind is a sensitive tooth. Anyone who has had one knows what sensitive in that context means. It does not take much to cause pain in that tooth, which is sensitive to everything. You may also know people who are sensitive in that their feelings are hurt easily. It does not take much to upset such people. That is similar to what we mean when we talk about the sensitivity of RF and microwave detectors. It is the input signal level required to make the detector operate and the amount of output voltage obtained for a certain amount of input power.

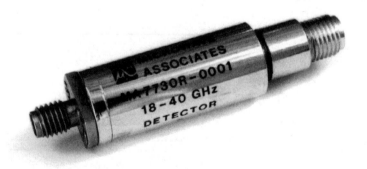

Figure 4.21 Detector.

Two factors affect sensitivity: the input power and the output voltage. Thus, sensitivity generally can be defined as how much voltage is produced at the output for a given input power level. Units such as millivolts (0.001V) per milliwatt (0.001W) are popular when we talk about the sensitivity of a detector. Another factor that plays a part in the sensitivity parameter is the load resistance, which causes the output voltage of a detector to decrease as the load resistance decreases. Thus, when specifying sensitivity, you must consider input power, output voltage, and load resistance. It does no good to specify the sensitivity for a detector by saying that it is 1,500 mV/mW, without also specifying at what load resistance the measurement was taken. So be sure to include the load resistance when you specify detector sensitivity.

Another type of sensitivity used with detectors is *tangential sensitivity* (TSS). A TSS display is shown in Figure 4.22. TSS is the lowest power level input that can be used to get an output signal that is above the noise of the device and allow for proper operation of the detector. TSS has been classified as a direct measure of the signal-to-noise voltage of the detector, since a certain noise is generated in the detector simply because there is a diode with a junction that exhibits a resistance. That is the most common noise source, a resistance with current flowing through it. As can be seen in Figure 4.22, the measurement is carried out with a pulsed signal. The pulsed signal is applied at the input to the detector, and its level is adjusted so the highest noise peaks observed on an oscilloscope with no signal are the same level as the lowest noise peaks when a signal is present. In other words, the presentation in Figure 4.22 is set with the input level until level B is equal to level A. When you get such a display, you can read the input power level in decibels referred to milliwatts and that will be the value of TSS. Typical values for TSS are −50, −55, or −60 dBm. This parameter is an important part of any detector data sheet.

Figure 4.23 shows the monitoring application of RF and microwave detectors. Notice that there are the two circuits (1 and 2) with the directional coupler between them. That part of it is exactly the same as in Figure 4.5, which showed a directional-coupler monitor circuit. The additional portion in Figure 4.5 was an indicating device connected directly to the directional coupler. In Figure 4.23, the detector is now

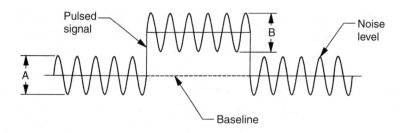

Figure 4.22   TSS display.

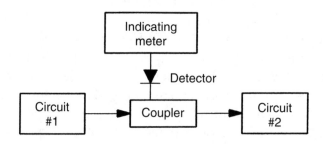

Figure 4.23   Monitoring circuit.

placed between the directional coupler and the indicating device. That means we can use a simple dc meter as an indicating device, rather than a power meter or a spectrum analyzer, which simplifies many applications.

Whenever we need to monitor a specific location in a system for specific parameters, the directional coupler and detector combination can be used to produce the voltage necessary. As a matter of fact, it is possible to purchase a *directional detector*, which is a directional coupler that operates over a certain band of frequencies and that has a built-in detector on it. The device has RF connectors at the input and the output of the coupler and a BNC connector at the output of the detector, a handy arrangement to have since it eliminates an additional adapter that would be used to connect the detector to the coupler. There is one thing to watch when you choose to use a directional detector: be sure it will operate over the frequency range you are using. When the two components are put together, it serves to decrease the operating band of the detector because the coupler usually has a limited range (e.g., an octave). When the detector is used as a separate component, the frequency range of operation can go from 500 MHz to upward of 18 GHz for one unit, depending on the connectors used at the RF port.

Another application of the RF and microwave detector is the leveling circuit shown in Figure 4.24. Leveling circuits were described in Section 4.1.2 and shown in Figure 4.6. The detector is the component that takes the sampled energy and converts it to a voltage that is used to control the voltage-controlled attenuator in the generator. The attenuator then has the task of adjusting the output level accordingly. This application many times also uses the directional detector. The directional detector often is used for a leveling circuit, because the generators may go over only an octave or two, and this component is ideal for such an application.

It can be seen from both Figures 4.6 and 4.24 how important the detector is to the operation of a leveling circuit. It has the responsibility of producing the "appropriate" voltage to tell the attenuator how much attenuation to place in the output line or take out. We emphasize appropriate, because it is a simple process to have a detector that will produce a voltage. The key is to have it produce the right voltage for the application you need.

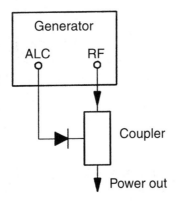

Figure 4.24 Leveling circuit.

A third application of the RF and microwave detector is shown in Figure 4.25. This application is for the times when no spectrum analyzer is available to look at an RF or microwave signal and you need to get an indication of what is going on in the circuit. If an oscilloscope is available with a limited bandwidth and frequency response, the detector can be used to provide a level indication for the signal that is being used. This application is being used less and less now because of the large bandwidths available in some of the new oscilloscopes.

An important part of the setup in Figure 4.25 is the attenuator (the block labeled ATTEN), which is inserted as a safety precaution. The attenuator's function is to limit the power that will reach the detector so there is no danger of burning out the diode and destroying the detector. It is a good idea to use such a device whenever you are working with an RF or microwave detector. First, you must look at the specifications for the detector to determine its sensitivity. Then the appropriate value of attenuator should be inserted to protect your investment. That saves time, money, and tempers later on, because the detector can be used over and over for many applications.

It should be evident that the detector is a highly reliable and simple component that has many critical jobs to perform in RF and microwave systems. The detector seems to bear out two old sayings: "Good things come in small packages" and "The simpler, the better."

4.5 Mixers

The concept of mixing is one that most people understand. When you make a cake (even from a box), you take specific materials, mix them together, and end up with a cake. If you mix different-color paints

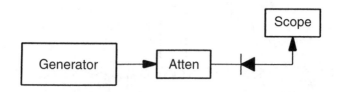

Figure 4.25 Oscilloscope.

together, you end up with a new color. Both these examples are rather basic, but they illustrate the process of mixing.

Similarly, when you mix together two electronic signals under the right conditions, you end up with a third, different signal (either the sum, the difference, or some combination of the two original signals). For both the cake and the paint, you need to observe certain conditions to make a presentable cake or get even close to the right color of paint. Similarly, certain conditions must be observed when we mix electronic signals.

If you put two signals into an amplifier, you get those two signals at the output (provided they are within the bandwidth of the amplifier and the levels of the signals will not overdrive the amplifier), because the ordinary amplifier is a linear device. That is, when you put an input to the device, you get a signal out that is multiplied by the gain of the device (a 1V input signal with a gain of 15 ideally results in a 15V output). Figure 4.26 is a curve of power input versus power output that shows this linear region. The area designated as linear is from the zero point at the lower left corner to the point marked A. It can be seen that this is a straight line with no bends in it or changes in direction. It also shows that for every increment of input power there is a corresponding increment in output power, which depends on the actual gain of the circuit being considered. As long as this characteristic is dominant in the circuit we are considering, there will be two output signals when we place two signals at the input.

When the linear relationship no longer exists, you have a nonlinear device, which also is shown in Figure 4.26. It is the area that comes about by increasing the input power level to a device until there is little or no increase in the output power. The point where that first occurs is called a compression point. It tells you, the operator, that you are leaving the linear region of the device and beginning to encounter the nonlinear properties of the circuit being used. When you get into a nonlinear area, you will notice that the input power continues to increase, with no increase in the output. There appears to be no increase in the output because what is being displayed is the amplitude of the fundamental, or original, signal. What actually is happening is that as the input level increases, the harmonics of the fundamental signal begin to increase at an ever increasing rate. That will cause harmonic distortion of the original

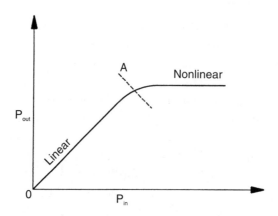

Figure 4.26   Power-in vs. power-out conditions.

signal, since the harmonics (which are two, three, four, etc., times the fundamental frequency) are now becoming so large that they are affecting the output of the device. That is actually nice for the operation of a mixer circuit since it is what is needed to get signals to mix together. As mentioned previously, two signals do not mix in a linear circuit. Each signal will be worked on individually by the linear circuit. Thus, there must be some method of getting the circuit into the nonlinear region that is needed for operation.

To have a mixer circuit, you need the nonlinear property of the circuit we have been talking about, but you also need to have the three fundamental parts shown in Figure 4.27: the input coupling network, the diode circuits, and an output filter. All three parts are absolutely necessary for a mixer circuit.

Before we get into the three sections of a mixer, it might be a good idea to describe the signals that are used as inputs and outputs of a mixer. As can be seen in Figure 4.27, there are two input signals, the RF and LO signals. The RF signal usually comes from a low-level source or directly off an antenna, and the LO controls the operation of the mixer itself. The output in Figure 4.27 is called the IF, which usually is the difference between the RF and the LO, but which also could be the sum of the two input signals. With the signals identified, we can now look at the elements that make up a mixer circuit.

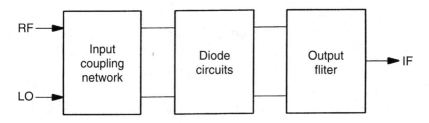

Figure 4.27 Mixer.

4.5.1 Elements of a mixer circuit

The input coupling network, as shown in Figure 4.27, is used to combine the input signals into the mixer, the RF and the LO. This network must provide a means of combining the two signals in the network and applying them to the diodes at the same amplitude. The RF and LO signals also have to be isolated from one another at the input, so that the only place they are combined is at the output of the coupling network at the input to the diodes. Recall that the hybrid junction in the quadrature hybrid provides excellent isolation between ports. It also combines signals and feeds them to the output at an equal amplitude. The quadrature hybrid, then, is an ideal choice for the input coupling network of a mixer. This, in fact, is the choice of many designers of RF and microwave mixers.

The diode circuits in Figure 4.27 can be a wide variety of assemblies: a pair of single diodes, an arrangement of four diodes in a quad (a bridge arrangement of four diodes) configuration, or any other combination that will do the job for a particular application. The main task of the diodes is to take the RF and LO signals and mix them together to form new signals. The main requirement is that there be an even number of diodes so that each side can be fed the combined signals evenly. (Some texts show a single-diode mixer configuration to illustrate the basic operation of the three sections that make up a mixer. Such a configuration is not recommended, since both the losses and the noise are very high, and you will not obtain optimum performance.)

The filter circuit at the output of the mixer is necessary because the result of mixing two signals in a nonlinear circuit is a wide variety of signals. If we let the RF signal be F_1 and the LO signal be F_2, the output of the diodes will be F_1, F_2, $F_1 + F_2$, $F_1 - F_2$, $2F_1 + F_2$, $F_1 + 2F_2$, and so on, with a large variety of signals produced as a result of the mixing process. Remember that this is a nonlinear operation and many signals will be generated during the operation. Actually, there is one big mess of signals at the output of the diodes, which is why a filter is required. If the difference between the RF and the LO signals is the required output of the mixer, a lowpass filter is placed after the diodes. That is because the lowpass filter passes the difference frequency and attenuates all the other frequencies since they are all at a higher frequency than the difference. If the sum of the RF and the LO signals is the desired output, the filter to be placed at the output of the diode circuits is a bandpass filter. That is because the frequency of the sum of the two signals is in an area that some of the signals will be higher in frequency and others lower. The filter to use, therefore, is a bandpass filter to ensure that you get the proper output signal. Figure 4.28 shows some mixers that are available.

4.5.2   Signals in a mixer circuit

For most applications of a mixer, the output is either the sum of the two signals ($F_1 + F_2$) or the difference ($F_1 - F_2$). If the sum is used, the mixer is called an *upconverter*. If the difference is used, the mixer is called a *downconverter*. The choice of an upconverter or downconverter depends simply on the type of filter that is used in the mixer. With the proper filter at the output of the mixer, you have a completed circuit. Now let us see how the signals that are used can make or break the mixer circuit.

Figure 4.28   Mixers.

We have said that the mixer must be a nonlinear device in order to mix the signals together. How do we get a circuit such as the one shown in Figure 4.27 to be nonlinear? For the amplifier example we looked at in the beginning of this section, we assumed that the signals were the same level and would be amplified equally in the linear case. For the mixer, the input signals are very different in level. The RF signal usually is a very low level signal. Many times it is the signal that comes in right off an antenna, and the level may be −70 to −80 dBm or even lower. The LO signal is the key signal for making the mixer circuit operate properly. The level of the LO signal is substantially higher than that of the RF signal and may be 0, +5, or as high as +27 dBm. This very high level hits the diodes at the same time the RF signal does and results in the diodes being driven very hard and into the nonlinear region. That is how the mixer creates the nonlinear effect and mixes the two signals. If the LO signal level falls off for any reason, there is a distinct possibility that the mixer will not mix the two signals. In that case, there will be no output from the mixer, since the output filter is designed to pass only the sum or difference frequency of the two inputs. So it can be seen that the level of the LO is crucial to the operation of the mixer.

Certain terms are associated with mixers. One term is *conversion loss*. Recall the term *insertion loss*, which describes directional couplers. Insertion loss is the low loss going from the input to output on the straight-through transmission line and usually is a very low value. The key idea behind insertion loss is that it is a loss from input to output of a device (e.g., the directional coupler). The conversion loss in a mixer is similar to the insertion loss in a directional coupler in that it is a measure of the performance from the input to the output. For a mixer, it is the difference in signal level (in decibels) between the RF input and the IF output. This loss is a little more difficult to measure than the insertion loss of a directional coupler because the input and the output of a mixer are at different frequencies, whereas the directional coupler insertion loss is at the same-frequency input or output. The best way to look at the conversion loss of a mixer is with a spectrum analyzer, which can distinguish between the two frequencies and give you an absolute power level for each signal. Then it is a matter of taking the difference in decibels and obtaining a figure for conversion loss. Typical values range from

6 to 9 dB. You also may see some mixer assemblies that do not have a loss from the RF input to the IF output. These are more appropriately called mixer-amplifier assemblies rather than simply mixers. An amplifier at the IF output raises the level of the output so there is a gain rather than a loss through the entire circuit.

Another term that is important when we discuss mixers is the isolation, particularly the LO-to-IF isolation. This value (in decibels) tells you how well the LO signal, which is considerably higher than any other signal in the mixer, is attenuated so it does not interfere with the desired output. Some mixer circuits have separate filters built in that are designed to greatly attenuate the LO signal at the IF output so that it does not overpower the required IF signal, thus producing a more efficient mixer circuit.

The value of RF-to-LO isolation also is important. It is necessary to keep the RF and LO signals apart at the input, and these signals should be combined only at the diodes so they can do the task they are supposed to do. Values for both RF-LO and LO-IF isolation should be in the range of at least 30 dB; 40 to 45 dB would be even better.

The main application for a mixer circuit is in a superhetrodyne receiver, which is shown in Figure 4.29. In that scheme, the RF signal comes in from an antenna (or possibly an RF amplifier right after the antenna) and is mixed with a LO signal to form an IF, which is used for processing the radio signal. The IF makes it much easier to process information, because it is a lower frequency and is also a single frequency

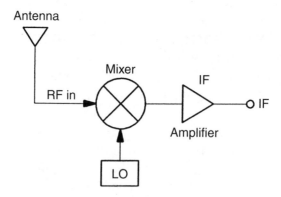

Figure 4.29 Superheterodyne receiver.

rather than a band of frequencies, like the RF input. Typical numbers for RF signals are 550 to 1600 kHz for commercial *amplitude modulated* (AM) radio applications and 88 to 108 MHz for *frequency modulated* (FM) radio cases. The IF frequency for AM is 455 kHz and 10.7 MHz for FM. Those frequencies would not be possible without the use of the mixer circuit to make the necessary frequency conversions. (A circle with an X inside is the recognized schematic symbol for a mixer circuit; see Figure 4.29.)

4.6 Attenuators

The term *attenuate* means to lessen the value of something. In RF and microwaves, attenuation is the lessening of the signal. Attenuating a signal in an electronic circuit is a matter of placing something in the path of the signal that will obstruct the signal's progress. That is fairly simple. However, to make an efficient attenuator that will operate as we would like it to requires that we not only attenuate the signal but maintain a good matched condition in the process. That is the measure of a good attenuator.

Different types of attenuators are used in RF and microwave applications. The fixed attenuator is probably the most common. Fixed attenuators are the components you see throughout a lab that are used for a variety of applications. Figure 4.30 shows two schematic representations that are used for fixed attenuators. Figure 4.30(a) is the T-type of attenuator, which has two series resistors and one shunt resistor. The π attenuator in Figure 4.30(b) is just the opposite, two shunt resistors and a single series resistor in between. Both of these fixed attenuators are very good for a variety of applications and are common in a laboratory environment.

The attenuators in Figure 4.30 can be used at a variety of frequencies. If your application is low frequency, a typical carbon resistor can be used. For higher frequency applications, ceramic chip resistors can be substituted for the carbon resistors. The substitution is a direct substitution, because a 3-dB attenuator to be used at 1 kHz uses the same resistor values as a 3-dB attenuator at 2 GHz. The only difference is the construction of the resistors: carbon versus ceramic chip.

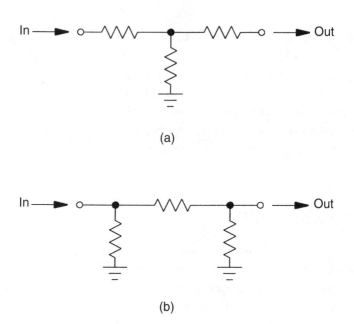

Figure 4.30 Fixed attenuators, (a) T attenuator and (b) π attenuator.

Fixed attenuators have, as the name implies, fixed values of attenuation, usually 3, 6, 10, and 20 dB. They may be as small as a little over 1 inch in length to 7 or 8 in, with heat fins for high-power applications. Fins indicate that an attenuator can handle much higher powers than a conventional fixed attenuator, which is an important consideration. Most of the fixed attenuators you see in a lab have a power rating of 1 or 2W. You should check the power rating closely when deciding which attenuator to use. If you do not, your lab may smell like burnt potatoes, which could get expensive, not to mention embarrassing.

Another type of attenuator is the variable attenuator. Two types of variable attenuators are used in RF and microwave applications: step attenuators and continuously variable attenuators. The step attenuator is a series of fixed attenuators that are switched in and out as needed to give the operator the selected value of attenuation. Or, as was mentioned in Section 4.2.3, the values of attenuation can be switched in by a digital pulse. The continuously variable attenuator is just what the name suggests: it continuously varies the attenuation to any value within the range of the attenuator. Recall that when we changed the bias on the PIN diodes that were placed between two quadrature hybrids we produced a continuously variable attenuator. The main point to bring out with the continuously variable or step attenuators is that not only are they good attenuators, but they also maintain an excellent VSWR as they provide this attenuation because of the constant-impedance structure of the devices. It also should be pointed out that not all attenuators are constructed with a pair of quadrature hybrids. Other attenuators use other means to maintain a good impedance match as they attenuate. Figure 4.31 shows a variable attenuator.

Some terms that apply to attenuators are VSWR, power rating, and attenuation. The VSWR is just as important as the value of attenuation we obtain. If the component only attenuates the signal and does not present a good VSWR, the attenuator is virtually of no use. Thus, a good attenuator will have a good VSWR over the full frequency band of operation. When measuring the input VSWR of an attenuator, it is important to terminate the output of the attenuator in its characteristic impedance. That will give you a good reading of what the device is really doing. Without the termination on the attenuator, you have an open circuit at the output and the only matching parameter is the value of

Figure 4.31 Variable attenuator.

attenuation going through the device. That gives you a number, but it will not be an accurate number for VSWR.

The power rating is important because if the rating is exceeded the attenuator will be degraded or destroyed. Thus, it is essential to note and observe the power rating of any attenuator being used, fixed or variable. That is particularly true if you have a higher power application. High-power attenuators are available; they usually have the characteristic black fins on them that are designed to dissipate the heat that high powers create. Regardless of your power requirements, it is a good idea to always check the power ratings on your attenuators (and all your components, for that matter).

Attenuation is a term that has been mentioned many times in this text. Attenuation is the value, in decibels, that tells you how much the signal level in your system will be decreased. As previously stated, fixed attenuators have typical values of 3, 6, 10, and 20 dB, with some attenuators being specially made at other nonstandard values. Step and variable attenuators may have any number of values of attenuation. Some step attenuators range from 0 to 1 dB in 0.1-dB steps; 1 to 12 dB in 1-dB steps; and 10 to 100 dB in 10-dB steps. Other step attenuators may be available upon request. Continuously variable attenuators have values in which the sky is the limit. Some go from 1 dB to 50 or 60 dB with no problem at all. To see all the possibilities, consult individual manufacturers catalogs.

Applications of attenuators are wide and varied. They are used to match circuits, reduce signal levels to detectors, and balance out transmission lines that otherwise would have different signal levels on them, to mention only a few applications. Any place where there has to be an adjustment to an RF or microwave signal level, the attenuator will be a necessary component. It should be pointed out that attenuators sometimes are overused. Some engineers tend to use attenuators as an easy fix for every problem that comes along. If, for example, it is difficult to match a particular circuit, some engineers will put an attenuator at the input and at the output and let it go at that. Doing that makes the circuit appear to have a good VSWR, but the circuit is still operating inefficiently and will not have the reliability it needs. Also, if a filter is causing problems in a circuit, placing an attenuator in series with that filter will make it look better, but the filter still is causing a problem. So, be careful when you are inserting attenuators in a circuit. They sometimes give you a false sense of security, and, as its name indicates, they do reduce the value of the signals you are using.

4.7 Filters

Filters have one basic function: to pass a particular thing and reject everything else. That is true of a car's oil filter (pass the oil and reject, or trap, dirt particles), a furnace filter (pass the warm air and trap the dust and dirt), or a pool filter (pass the water and trap dirt, leaves, twigs). Each of those filters provides a clean operation in some form; all are filters we encounter every day. The idea behind those filters also applies to RF and microwave filters: provide a clean signal for the system. By "clean" signal, we mean the one and only desired signal or band of signals required for the system's proper operation. The filters reduce or eliminate spurious signals or harmonics.

As we will show, filters are an important part of RF and microwave systems. They are, however, often overused because they do such an excellent job of cleaning up the signals. It must be understood that whenever you add a filter to a circuit, you are adding loss, VSWR from the filter, ripple (described in Section 4.7.1), and even some delay in the circuit. So, the use of filters should be considered carefully; in some cases, there may be an alternative means of cleaning up the circuit or system.

This section covers three types of RF and microwave filters: bandpass, lowpass, and highpass filters. The response curve of each type of filter is presented, in order to introduce the filters as separate components. The common terms associated with each filter also are presented, along with typical microstrip and stripline representations (the common methods of construction that apply to RF and microwave filter circuits). Finally, applications of the filters are discussed.

4.7.1 Bandpass filters

A bandpass filter does exactly what its name says: It passes a specific band of frequencies and rejects frequencies below and above that band. The response curve for a bandpass filter is shown in Figure 4.32. This response shows an area that is termed the *passband*, which is the area where there is a minimum loss in the filter response. On either side of the passband, you can see that the signal is greatly attenuated. Those dropoffs are called the filter skirts and fall off at a rate determined by the attenuation as a function of frequency. For example, if we have an attenuation of 50 or 60 dB a few MHz away from the passband, the skirts would be very sharp. If, however, the attenuation is only 15 or 20 dB, there would be a much gentler slope to the skirts.

Another term shown in Figure 4.32 is the *insertion loss*, which is the internal loss through the filter within the passband. It is very much like the insertion loss in a directional coupler in that it is the minimum straight-through loss of the device. No device will have 0-dB insertion loss, simply because there is energy passing down a transmission line. Therefore, a small loss in the passband depends on the width of that passband. The passband is determined by the number of sections in the

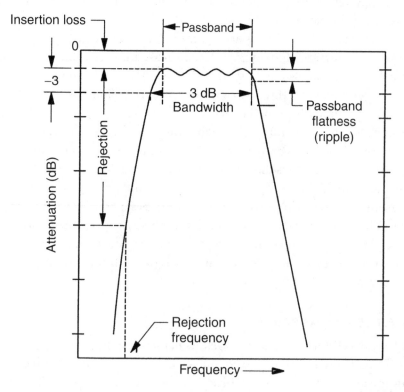

Figure 4.32 Bandpass filter response.

filter and how they are coupled. If the sections are loosely coupled, the passband will be wide. Conversely, if the sections are tightly coupled, the passband will be very narrow. The coupling of sections is accomplished in bandpass filters by the spacings between the microstrip or stripline sections.

The *ripple* of a filter (or *passband flatness*) comes about as a result of the number of sections (or poles) used to construct the finished filter. Each pole has its own response curve. Each response curve has a specific center, or resonant, frequency of operation. They are actually separate filters themselves with response curves similar to the curve shown in Figure 4.32 and which can be seen in Figure 4.33. The top portion of Figure 4.33 shows three sections of a filter and how each section has its own individual response curve: f_1, f_2, and f_3. The result is shown in the lower portion of the figure. The three responses are shown coupled together to form the total response curve for the filter. The coupling of these sections, which is not perfect, results in variations in the passband; thus, a ripple is produced in the passband. You should also be able to see that if we very loosely couple the sections together, the ripple will be very large. Similarly, tightly coupled sections reduce the ripple and also reduce the passband greatly. So the coupling factor of each individual section of a filter is critical both to maintain the passband desired and to

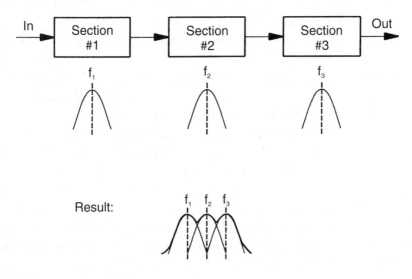

Figure 4.33 Section response.

keep the ripple under control. One thing that is not needed in a filter is a passband that varies greatly in amplitude.

The bandwidth shown in Figure 4.32 is characterized as the 3-dB bandwidth of the filter. That is established by the fact that it is the resulting bandwidth of the filter where the response curve is decreased by 3 dB from the insertion loss of the filter. That is, a generator is placed at the input to the filter, and a power meter, or spectrum analyzer, is placed at the output. The passband of the filter is found by locating the lowest attenuation point of the filter (insertion loss). The frequency of the generator then is increased until the power level drops 3 dB from the insertion loss, termed f_{upper}. The generator frequency then is decreased back through the passband and down until the response drops 3 dB on the lower side of the response curve, f_{lower}. If we now take $f_{upper} - f_{lower}$, we will have the 3-dB bandwidth of the filter.

The bandwidth of a filter does not need to be the 3-dB bandwidth. There may be other values: 1 dB, 2 dB, ripple bandwidth, 5 dB, and so on. It all depends on the particular application. Whatever the application is, that is what the bandwidth should be. The bandwidth parameter supports the idea that there is no such thing as a "standard" filter. You cannot just go to a catalog and find the exact filter for your application. The center frequency may be right, but probably not too many more parameters will match. The majority of filters are special filters rather than off-the-shelf standard filters that everyone can use.

The term *rejection* tells how much an undesired frequency is attenuated on the skirts of the filter. It can be seen from Figure 4.32 that the rejection point has two values: a decibel figure for how far down the signal should be and a frequency at which the rejection is to take place. You cannot specify rejection simply as 60 dB, for example. You need to have a specific frequency associated with the attenuation figure. Values of rejection can be specified both above and below the passband. That is because a bandpass filter is used when there is only a small band of frequencies being passed compared to the entire RF and microwave spectrum. It is logical, therefore, to have certain frequencies both above and below the passband that you do not want to interfere with operations of your circuit. Some typical values for rejection are 50 to 60 dB. The value, of course, depends on the individual requirements of the application. There may be instances where you will need a frequency attenuated

70 dB, while other times a simple 20-dB rejection of a frequency will do the job nicely. Be sure to know the requirements before specifying the rejection. You will save a lot of money if you do not overspecify your filters.

A variety of construction techniques can be used to fabricate bandpass filters in either microstrip or stripline. Figure 4.34 shows three types of bandpass filters. The first, a side-coupled half-wave resonator filter, is fairly easy to fabricate. It has a series of resonators side by side with designed gaps between them for coupling purposes to obtain the final response. This filter shows very well the idea of having a series of sections, or resonators, that are coupled together simply by placing them close to one another. One problem that may arise with this type of filter is that it may end up being very long and narrow when many sections of filter are used. When that happens, the filter can be "folded" so it is not as long. Folding is an acceptable procedure and makes the filter much easier to handle.

The second type of bandpass filter in Figure 4.34 is a short-circuited quarter-wave stub filter. In this type of filter, all the elements are a quarter-wavelength long. The ends of the extended resonators are shorted to ground, which results in the proper bandpass response. It can be seen that this filter is not a coupled type of filter but relies more on the properties of transmission lines and the fact that a shorted quarter-wave line will appear as an open circuit on the other end. An interesting point about this filter is that as it appears in Figure 4.34 it is a bandpass filter. If you were to remove all the ground connections and leave them open, you would have a band-reject filter, that is, the filter would reject the exact band that it originally passed. This is because of the transmission line characteristics covered in Chapter 3.

The third type of bandpass filter shown in Figure 4.34 is an interdigital filter. This type of filter has a series of quarter-wave resonators grounded at alternate ends to form the required bandpass response. For many years, interdigital filters were difficult filter to fabricate because it was almost impossible to obtain a good ground connection at the ends of the resonators. It would be much easier if the grounds were all on the same end. But, as can be seen in the figure, they alternate ends. Today, plated through holes can be fabricated that result in excellent ground connections.

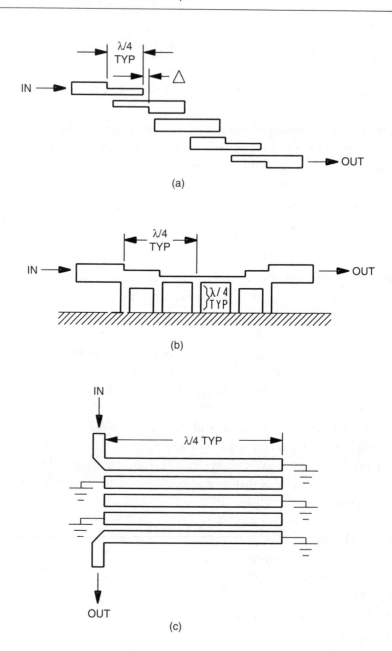

Figure 4.34   Bandpass filters: (a) side-coupled half-wave resonator bandpass filter, (b) short-circuited quarter-wave stub bandpass filter, and (c) interdigital bandpass filter.

Bandpass filters find applications in areas where they can be used to reduce noise and harmonic content of a system by limiting the bandwidth of the signals seen by that system. They also can be placed at the input to a receiver for improved selectivity. Many times bandpass filters are used in a system to "clean up" the system response. Spurious signals may be creeping into the system or excessive levels of harmonics may be causing problems with the required band of frequencies. Bandpass filters often are used at the outputs of signal generators in the laboratory. They usually are there as a precaution so you do not have to worry about the purity of the signals being sent to a test setup. Figure 4.35 shows a bandpass filter.

4.7.2 Lowpass filters

Another type of filter is the lowpass filter. Figure 4.36 shows the response of such a filter. As its name implies, a lowpass filter passes frequencies below a certain frequency with very little loss and attenuates frequencies above that frequency. This *cutoff frequency*, f_c, is the point where the response curve falls 3 dB below the insertion loss of the filter. You must

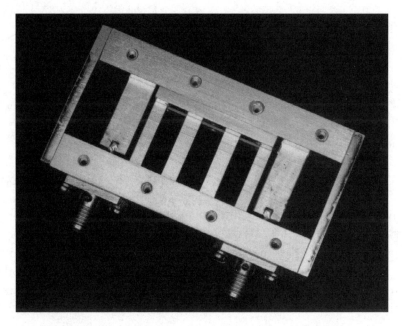

Figure 4.35 Bandpass filter.

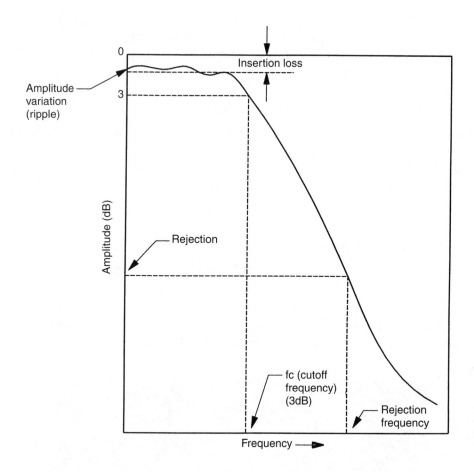

Figure 4.36 Lowpass filter response.

be careful when specifying a lowpass filter that you do not take the highest frequency you want to pass and call out that type of lowpass filter. If, for example, you wanted to pass everything below 900 MHz, specifying a lowpass filter with f_c = 900 MHz would result in the 900 MHz signal being attenuated 3 dB. To pass the 900 MHz, you would need to specify a filter with an f_c of 1.0 or 1.1 GHz (1000 MHz or 1100 MHz) to be safe. Another point about lowpass filters is that they pass dc. Their equivalent circuit is a series inductance with a parallel capacitance, as can be seen in Figure 4.37. From the figure it is easy to see how a lowpass filter can pass dc with nothing but a low-resistance coil between the input and the output. When you stop and think that a coil is nothing more than loops of regular wire, you can understand how it will pass dc easily. The reason this is pointed out is that lowpass filters can be used with amplifiers or circuits that require a dc voltage to operate. A filter placed in the wrong place may divert the dc from its required location to some other point where you do not want dc. So remember that a lowpass filter will pass dc and act accordingly.

A lowpass filter can also have rejection points. As with bandpass filters, lowpass filters can have specific frequencies called out with rejection values (1 GHz at 40 dB, 2 GHz at 60 dB, etc.). Once again, the specific points and attenuation values depend on the particular application. The number of rejection points also is controlled by the application.

The common nomenclature for a lowpass filter shown in Figure 4.37 can be used with either microstrip or stripline. The figure shows the narrow lines as series inductance and the wide pads as the top plate of the parallel (shunt) capacitance. (You can relate this to Figure 4.38, which is the equivalent circuit of a lowpass filter and which shows series inductors and shunt capacitors. The discussion here is how those elements are fabricated in stripline or microstrip.) The bottom plate of the capacitor is the ground plane underneath the circuit. A dielectric material separates the two plates, which completes the structure of a shunt capacitor to ground. This very good lowpass filter design is relatively easy to fabricate and reproduce in large quantities. It is probably the most common type

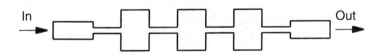

Figure 4.37 Lowpass filter.

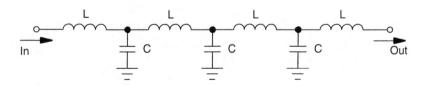

Figure 4.38 Equivalent circuit of a lowpass filter.

of lowpass filter for RF and microwave applications. Figure 4.39 shows lowpass filters that are available.

One application of lowpass filters, discussed in Section 4.5, is to provide the difference between the RF and LO inputs at the output of a mixer. A lowpass filter is used because all the other frequencies generated by the mixer will be higher in frequency than the difference component. A lowpass filter is the most logical choice for that application.

A second application of lowpass filters is the removal of spurious signals and harmonics from generators or systems in which too many signals may be present. Many times, a lowpass filter is placed at the output of a generator or source that is rich in harmonics, that is, the output of the generator or source contains many harmonics that are at a higher level than can be tolerated. The lowpass filter is ideal for the removal of these unwanted pests.

4.7.3 Highpass filters

The highpass filter is exactly the opposite of the lowpass filter, as Figure 4.40 shows. A highpass filter passes frequencies above a specific cutoff frequency, f_c, which is also 3 dB below the insertion loss of the filter, just like the lowpass filter. Highpass filters also have insertion losses and

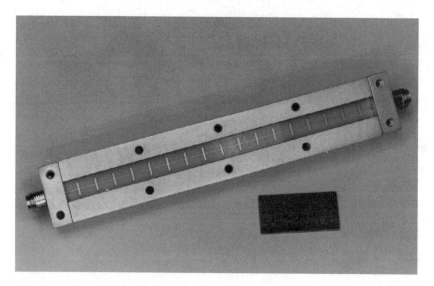

Figure 4.39 Lowpass filters.

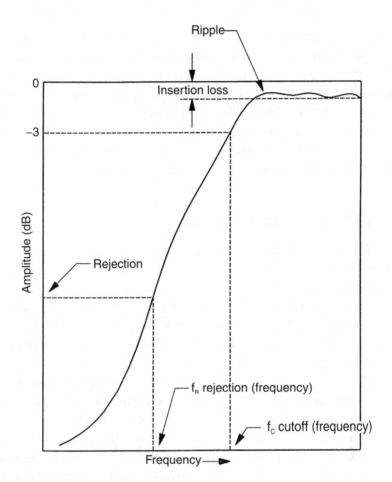

Figure 4.40 Highpass filter response.

certain specified rejection frequencies. A highpass filter will not pass frequencies up to blue light. It will have a low loss up to a certain point but will start to pick up attenuation as the frequency gets higher and higher. There actually is an upper limit to a highpass filter, but it is well beyond the frequency range where any highpass filter would be designed and is not cause for concern.

Highpass filter do not have as many applications as lowpass and bandpass filters. One of the main applications is when low-frequency signals cause problems in a system. A highpass filter will attenuate those signals and help the system operate properly. It also may help when a test generator has a high degree of low-frequency signals. Along those lines, highpass filters are used in some RF and microwave transistor circuits to reduce the low-frequency gain that some high-frequency transistors have. There may be times when a high-frequency amplifier looks unstable but does not show any signs of oscillating. A closer look at the circuit reveals that there is a tremendous amount of gain at a very low frequency and the circuit is actually oscillating at that low frequency. A highpass filter will reduce that gain and keep the device from oscillating.

4.8 Circulators and isolators

Section 4.2 discussed the directional coupler, a component that has directional properties and no dc connection. The coupling takes place across a gap between the transmission lines, and the directional property comes about because of the length of the coupling area. This section discusses components that also have directional properties, but this time they have complete dc continuity. These components are the circulator and the isolator (the isolator is a special case of a circulator).

The directional property of the coupler comes about by the use of specific-length transmission lines, resulting in the device exhibiting a quarter-wave response at the appropriate frequency. With the circulator, there is an interaction between a magnetic field and a ferrite device that causes a gyromagnetic motion and results in highly directional properties for the component. To understand this gyromagnetic motion, think of a pail of water into which you place a paddle and begin stirring in a clockwise direction. If you drop some wood chips into the pail as you continue to stir, you will notice that the wood chips easily follow the circular motion of the water in the clockwise direction. Now imagine that you want some wood chips to go in a counterclockwise direction. You would readily see that that would be impossible because the water motion is too strong in the clockwise direction. The same thing occurs with a gyromagnetic motion in a circulator. The interaction of a magnetic field and the ferrite material inside the circulator creates a circular motion of the magnetic field. The motion can be either clockwise or counterclockwise, depending on the orientation of the magnets. The rotating magnetic field can be very strong and will cause any RF or microwave signal placed at one port to follow the magnetic field around to another port and not be able to go in the opposite direction. Thus, we again have a directional device but with an entirely different set of circumstances that create the directional property.

To further understand what is happening in a circulator, look at Figure 4.41, which illustrates the construction of a circulator. First, look at the circuit board in the center of the device in Figure 4.41. Note that a circuit is printed on the board that is three lines connected in the center. That is where the dc continuity comes in. If you were to take an ohmmeter and go between ports on a circulator, you would have complete continuity. That must be taken into consideration when you use circuits that require a dc bias. Recall that dc must be considered in lowpass filters, too, for the same reason. The small transmission lines that come off the

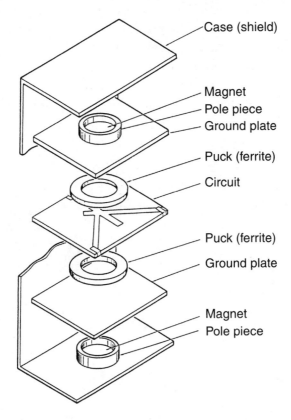

Case (shield)

Magnet
Pole piece
Ground plate

Puck (ferrite)

Circuit

Puck (ferrite)

Ground plate

Magnet
Pole piece

Figure 4.41 Circulator construction.

circuit in the center of the board are for matching purposes. Because the transmission lines are just stuck together in the center, there will be a mismatch at the junction. To maintain a good match for the circulator, there needs to be some method to eliminate the mismatch at the center of the transmission lines.

On either side of the circuit board is a *puck*, which is the ferrite material with which the magnetic field will interact. (The ferrite material is commonly referred to as a puck because it looks like a hockey puck.) The next layer of the circulator is a ground plane board, which makes it a very nice stripline package.

On either side of the ground plane board is a magnet (pole piece), which, of course, is the source for the magnetic field that interacts with the ferrite device to make the gyromagnetic motion. The package is completed by a shield around the entire circuit to keep the magnetic field of the component from interfering with other circuits in the area. The shield also keeps the magnetic field inside the circulator so it can totally interact with the ferrite within the component.

Figure 4.42 shows the schematic representations for a circulator and an isolator. Note that the isolator is actually a circulator with the third port terminated in the characteristic impedance of the system. Both

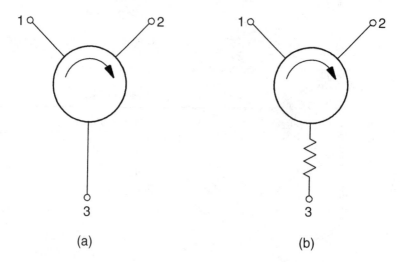

(a)   (b)

Figure 4.42   (a) Circulator and (b) isolator schematics.

devices can be used for a variety of applications in RF and microwave circuits and systems. The arrows show the direction of the magnetic fields and will aid you in telling which way the energy will flow when it is applied to the circulator or the isolator. Figure 4.43 shows circulators that are available.

To further understand how the circulator and the isolator have many applications, let us go through the basic operation of the devices. If the input signal is placed at port 1 of the circulator in Figure 4.42, it will follow the arrow, which represents the interaction of the magnetic field and the ferrite device to produce a strong field in the direction shown. If the load at port 2 is fairly well matched, the signal will exit at that port with very little loss from the input level, usually in the order of 0.2 to 0.5 dB. If there is a mismatch at port 2, the signal will be reflected back from port 2 and be sent to port 3. If there is a match, the signal will exit here; if not, it will be reflected. The signal will be reflected back to port 1, and you will accomplish nothing but inserting an expensive wire in the line.

With the circulator operation in mind, it is a straightforward step to see how the isolator works. If we still use the input as port 1, the output will be at port 2 if the load is matched. If there is a mismatch at port 2,

Figure 4.43 Ferrite circulators.

the signal will be reflected from port 2, follow the magnetic field around, and be deposited into the termination of the device. Insertion losses for the circulator and the isolator are on the order of 0.5 dB with isolation figures in excess of 20 dB. Those values are general numbers that can vary from unit to unit. The amount of isolation present in the isolator depends heavily on the termination at the third port. If the match at that port is on the order of 2:1, the best isolation that can be expected is approximately 9 dB. If the match is improved to 1.5:1, the isolation increases to 14 dB. If we get a really good termination and the match is on the order of 1.05:1, the isolation will be 30 dB. Thus, it can be seen that it is important to have a good termination on the isolator so it can do its job. That, by the way, is also the case for the circulator. If you want the circulator to separate signals as it is intended to do, the components that are placed on each port must exhibit a good match.

Figure 4.44 shows two of the many applications for a circulator. Figure 4.44(a) shows a duplexer, which is used when a transmitter and a receiver have to use the same antenna. As can be seen from the figure, the transmitter is placed at port 1, the antenna at port 2, and the receiver at port 3. When the transmitter sends out a signal, the signal goes to the antenna with great ease and does not leak into the receiver because of the isolation of the circulator. This is one area in which there must be very good isolation between ports to ensure that the high-power transmitter output does not get into the receiver and destroy the front end. When the signal comes back to the antenna, it goes directly to the receiver and not to the transmitter, because of the circulator operation. It should be evident that to have the required isolation in the circulator the transmitter, the receiver, and the antenna all must be well matched to the circulator. If that is not accomplished, the duplexer will not perform as expected.

The application in Figure 4.44(b) is used with a tunnel diode to produce an amplifier. The circulator is placed in the center of the amplifier section and it applies the input signal to the amplifier. Because the tunnel diode is a negative resistance device, there are a great many reflections from it. The reflections are sent back through the circulator to the output. This application was a commonplace one for circulators for many years for use in satellite application since tunnel diode amplifiers

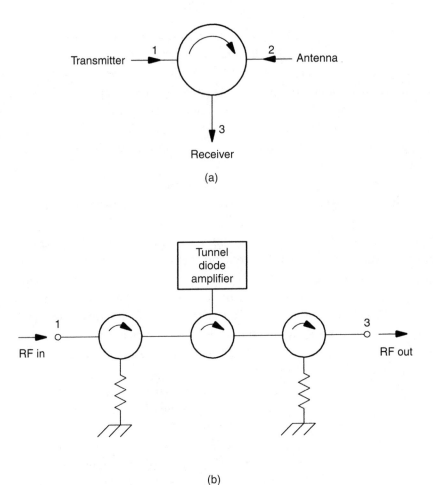

Figure 4.44 Circulator applications: (a) duplexer and (b) circulator with tunnel diode amplifier.

are very low power devices. The application has decreased in recent years, but many of these circuits are still around.

Figure 4.45 shows three applications for isolators. The application in Figure 4.45(a) is designed to keep the high level of an LO in a mixer circuit in the mixer itself and not have it radiate out through the incoming antenna. Because, as was discussed in Section 4.5, the LO power level is much higher than the RF signal that is coming in from the antenna, there is the possibility of this signal leaking back to the input circuit. That circuit may be attached to an antenna, as shown in the Figure 4.45(a), and the LO signal will radiate out into the air. That is possible because the RF and LO signals are not that far apart in frequency, and the antenna usually will very easily transmit that signal out into the air. With the isolator in the circuit as shown, the signal coming from the mixer will hit the isolator and be dropped instantly into the termination. That keeps the signal in the mixer, where it can do the job it was intended to do.

The application shown in Figure 4.45(b) is designed to supply a constant load to an oscillator circuit. If the load attached to an oscillator varies in value, there is a good possibility that the oscillator can be pulled off frequency and have a different output level than it was designed to produce. With the isolator in the circuit, variations in the load will be sent back to the isolator and end up in the termination, where the oscillator will never see them. This application is used many times when you have a transmission system in which the carrier frequency and level must be held constant. The isolator is a small, relatively inexpensive way of ensuring that these properties are preserved.

The application in Figure 4.45(c) is similar to the second one. In this application, the isolator is placed between the generator and a test setup. Generally, when a circuit or system is being tested, there may be variations in the test setup that you do not want the generator to see. If the isolator is used, the variations are sent to the termination of the isolator and never get to the generator. (Figure 4.46 shows an isolator).

4.9 Antennas

When the term *antenna* is mentioned, many people think of a car antenna, either a long metal rod or a piece of wire imbedded in the windshield that allows them to hear their favorite radio station. That, of course, is one application of antennas, but there is much more to an antenna than a piece of metal sticking up in the air. Antennas are a vital part of many RF, microwave, and wireless applications, both commercial and military.

The simplest, and probably the most understandable, way to explain antennas is to start with a transmission line with its output end left open.

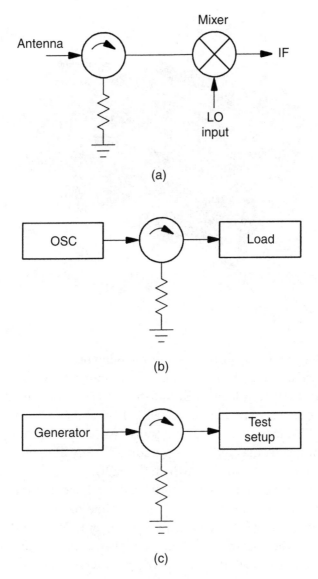

Figure 4.45 Isolator applications: (a) local oscillator radiation reduction, (b) oscillator pulling reduction, and (c) generator isolation.

You will recall from the discussion of open-circuited transmission lines that the voltage at this point is a maximum value, and there will be radiation of the energy from the open-ended line. That is something that

Figure 4.46 Ferrite isolators.

will happen even though you are not planning on making your transmission line an antenna. Any open-ended transmission lines, including microstrip transmission lines, will radiate and act as antennas. That is the reason these lines are kept a distance from other transmission lines. Figure 4.47 shows an open-ended transmission line. Although a certain amount of electromagnetic energy radiates from all open-circuit transmission lines, the distance for which the radiation is of any consequence is minimal. That certainly would not be a very good intentional antenna. What we need is some method to get the line to radiate power over a much longer range. We could, of course, put a tremendous amount of power in the transmission line, but that is impractical and costly. Figure 4.48 shows a good start in getting more range out of the transmission-line antenna. The open transmission line now has a flared end on it. That allows more energy to be radiated out into the air. The flared structure is the beginning of a familiar structure called a dipole. When we talk about a dipole antenna, we mean that the antenna has two poles associated with it. One is on the top conductor, and the other is on the bottom conductor. The combination of the two poles makes an economical and efficient antenna.

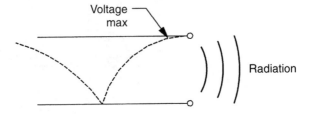

Figure 4.47 Open-ended transmission line.

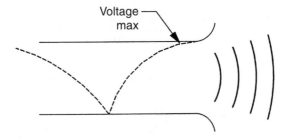

Figure 4.48 Start of a dipole antenna.

Figure 4.48 is the beginning of the dipole antenna because the antenna usually does not look like the flared transmission line in the figure. It is presented like this to lead you into the actual construction of the common dipole antenna. The more recognizable representation of a dipole antenna is shown in Figure 4.49, which is a picture of the initial dipole antenna shown in Figure 4.48 with the conductors expanded until the distance between them is a quarter-wavelength ($\lambda/4$). This type of antenna is called a quarter-wave dipole or a Marconi antenna. (Keep in mind that the drawings are not to scale. The distance between the transmission line conductors is actually very small compared to the actual length of the antennas they are creating.)

Figure 4.50 shows the most popular type of dipole antenna, which is the same type of antenna just described except that we have continued to spread out the transmission line until each side is a quarter-wavelength. It now becomes a half-wave dipole antenna, commonly called a Hertz antenna. (Even though it is called a half-wave antenna, the length of the

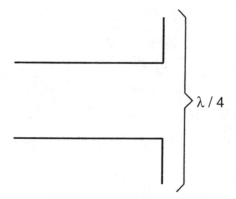

Figure 4.49   Quarter-wave dipole antenna.

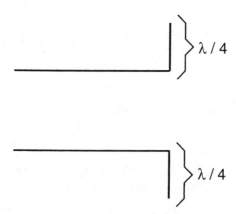

Figure 4.50   Half-wave dipole antenna.

antenna usually is decreased by about 5%. When that is done, the wavelength that is actually used is 0.48 wavelength rather than $\lambda/2$. If that relationship is used, the antenna most likely will work exactly on the frequency you need.) Figure 4.51 shows an available antenna.

Keeping in mind how a transmission line can be changed into an antenna, let us now look a little deeper into antennas. The first concept to understand about antennas is that they are reciprocal devices, that is, they can be used for both transmission and reception. Think of a CB radio where a single antenna is used for both talking and listening. Therefore,

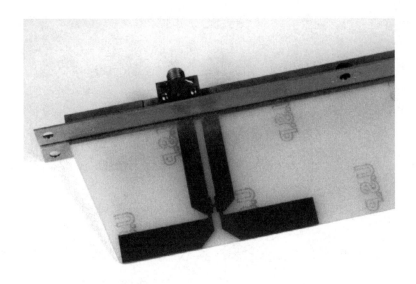

Figure 4.51 Dipole antenna.

the antenna must be reciprocal and be as efficient one way as it is the other.

To further understand antennas, you must know the terminology used to describe them. One term used for all antennas is the *radiation pattern*, which is a plot of the strength of the radiation being sent out (or the area where it can be received) as a function of the angles around the antenna. Figure 4.52 shows a typical radiation pattern for an antenna. The majority of the energy is directed out in one direction. That is called the front lobe and is considered to be the major lobe. The front lobe is actually what you look at when you evaluate an antenna. (If the antenna is an omnidirectional antenna, the energy will be distributed over the entire 360-degree area uniformly.) Figure 4.52 also shows some smaller lobes on the sides of the major lobe. These side lobes are classified as minor lobes and are the result of a certain amount of leakage out the sides of the antenna. Also, a certain amount of energy leaks out the back of the antenna. That is the back lobe, which also is classified as a minor lobe. The radiation pattern is a sort of "fingerprint" for an antenna. It identifies the antenna and its characteristics.

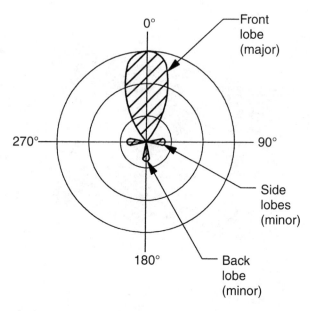

Figure 4.52 Radiation pattern.

One way of identifying the properties of an antenna is to use the concept of near fields and far fields. The near field is the field generated very close to the antenna. Near fields sometimes are called induction fields because of the induction action that takes place close to an antenna. This is similar to current flowing through a conductor and creating a magnetic field around the conductor, which is the basic idea of motor theory. A far field is defined as anything beyond the near field. Once the energy reaches the far field, it continues to radiate out into space for some distance. For that reason, it often is called the radiation field. The radiation in the far field never returns to the antenna, as some of the near field energy does.

An important term to understand with antennas is *effective isotropic radiated power* (EIRP), which is the actual power radiated from the antenna. Most of the time when transmitter power is mentioned, it is thought that that is the power sent out into space. Actually, a lot more power goes out into space than comes from the transmitter. To understand this concept, you must understand that antennas have gain. Gain makes the power coming out the antenna much higher than simply that of the transmitter output. The EIRP is the product of the transmitter power and the antenna gain. Consider the following example. A transmitter has a power output of 10W (+40 dBm). The antenna we will use has a 30-dB gain, which means the actual output power from the antenna is +70 dBm, or 10,000W. That amount is much higher than the original 10W coming from the transmitter and can cause problems if you do not consider the antenna gain and use EIRP on all your power calculations.

A special type of antenna that finds many applications in RF, microwave, and wireless systems is the patch antenna, two types of which are shown in Figure 4.53. This type of antenna is a microstrip construction that uses high-frequency material with a complete ground plane on the reverse side. It is a square patch (sometimes round) etched on a circuit board. The feedpoint (where the signal is sent to the antenna) can be either at an off-center point of the patch, as shown in the top drawing in Figure 4.53, or at the edge of the board, as shown in the bottom drawing.

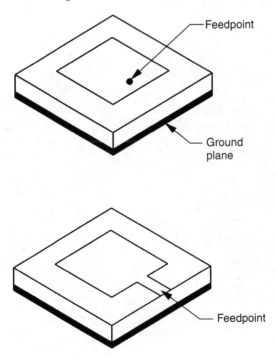

Figure 4.53 Patch antenna.

When the patch antenna is a square, each side is a half-wavelength long. When the patch is circular, the antenna works best if the diameter is 0.6 wavelength across. (Figure 4.54 shows the patch antenna.)

Another antenna that is used for wireless communications is a monopole antenna. This antenna is a dipole with one-half of it replaced by what is called an infinite ground plane, a very large ground plane that creates an image of the dipole piece that is used. The monopole antenna has a length of a quarter-wavelength.

The loop antenna may be either a circular or a square configuration. These antennas are self-resonant when the outer perimeter is equal to one full wavelength at the frequency of operation. For example, the circumference of a circular loop or the outer perimeter of a square loop would need to be approximately 34 cm (13.38 in) to operate at 900 MHz. This type of antenna gets a bit unruly when the frequency is low. Consider a loop antenna at 150 MHz, compared to the 900 MHz example. That antenna would have a circumference (or outside perimeter) of 204 cm, or 80.31 in, which would be a circular antenna with a diameter of

Figure 4.54   Patch antenna.

25.56 in or a square antenna 20.07 in across. Contrast that to the 900 MHz model that would have a circular diameter of 4.25 in and a square antenna that would be only 3.34 in across. These dimensions lend themselves much better to wireless applications.

4.10 Summary

This chapter described the components that make up RF, wireless, and microwave circuits that make up some very sophisticated systems. It covered directional couplers, quadrature hybrids, power dividers, detectors, mixers, attenuators, filters, circulators and isolators, and antennas. The operation of each type device was explained, terms associated with them defined, and examples of each presented. These components truly are the main parts of RF and microwave systems.

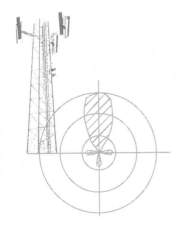

5

Solid State Devices

SOLID STATE DEVICES in RF and microwave are an integral part of the commercial advancements that have taken place over the past few years. The entire field of wireless communications would not be possible without these devices. If you look back at the first devices used to amplify signals and produce such components as oscillators, diode detectors, and signal modulators, you can see that the miniaturization that has taken place over the past few years would be impossible without solid state devices. The early days of electronics used vacuum tubes, which were large and bulky and needed an ac voltage to heat the filaments so electrons could flow within the device. Let us briefly look back at the vacuum tube and how it evolved into the transistors and diodes that make up the world of RF, microwave, and wireless markets today.

Look through any electronic books from the 1950s or 1960s, and you probably will see the symbols for vacuum tube devices. You also may wonder just what these things were and how they possibly could work. Figure 5.1 shows two types of vacuum tubes: the diode and the triode.

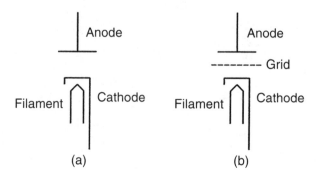

Figure 5.1 Vacuum tubes: (a) diode and (b) triode.

The diode tube has a filament and two elements: the cathode and the anode. The cathode is the source of electrons that are sent to the anode (sometimes called a plate). A diode operates as follows. An ac voltage (6.3V or 12.6V) is applied to the filament of the tube. (The filament is the portion of the vacuum tube that is glowing.) The filament is heated by the ac current that flows through it. That causes the free electrons on the metallic cathode to be "boiled off" and form a cloud of electrons just off the surface of the cathode. The cloud remains there if the diode is not turned on. To turn the device on, we apply a positive voltage to the anode element of the tube. The voltage usually is in the range of +150V to +200V. By using some basic reasoning, we can see that the large positive voltage will attract the electrons from the cathode to the anode. The result is a current flow through the device. Because this is a diode, the current will flow in one direction. Diode tubes, which were used for many applications in early circuits, can be used as detectors, half-wave rectifiers, and full-wave rectifiers for power supply circuits.

The circuit in Figure 5.1(b) is a triode. If you compare the triode to the diode, you will see that there is an additional element in this device, a grid. The grid is a wire mesh that is used to control the amount of current that flows in the device. It is like a valve that determines the number of electrons that will get to the plate of the tube. The grid has the input signal to a device placed on it. For example, consider a sine wave applied to the grid and what it does for the operation of the triode tube. When the input signal goes positive, the grid attracts electrons from the cathode to be sent to the anode. When the grid reaches its peak positive value, the maximum amount of current will be flowing. As the signal goes beyond this maximum point, the signal will go less positive, and the current will begin to decrease. When the input signal reaches its maximum negative value, there will be a minimum current flow through the device. The process continues, and the input signal is reproduced at a higher level at the output of the tube by a value determined by the gain of the circuit.

Thus, the vacuum tube has a very logical operation that depends on simple positive and negative relationships and how like charges repel each other and opposite charges attract. Also, the third element, the grid, has the important task of providing control of the current through a basic diode circuit. These concepts are presented here because the operation of a solid state device is not much different, in theory, from the vacuum tube. There are, however, no filaments that need to be heated, and there is no need for large voltages to operate the solid state devices.

With what the early days of electronics had to offer in active devices (active devices require a voltage to operate), let us now investigate modern solid state devices and how they relate to RF and microwave applications. We will look at a variety of devices used in RF and micro-wave circuits, as well as the terms that are used to describe them.

The first order of business is to present and define some common terms used in solid state technology. The first term we will define is one that often is used without any idea of what it actually means: *bipolar*. All some people know is that bipolar refers to some sort of transistor that does a nice job for them. The term is, in fact, associated with transistors that can be used for RF and microwave applications. To understand this term, you first must realize the meaning of the prefix *bi*. You may recall that in 1976 the United States celebrated its 200th birthday, or its bicentennial. The Olympic Games include the biathlon, which involves a combination of two events, cross-country skiing and rifle sharpshoot-ing. A person who can speak two languages fluently is said to be bilingual. You can see where we are going. The prefix *bi* means two, which is also how it relates to the term *bipolar* in RF and microwave applications.

Most devices in electronics rely on a single means of moving energy through that device, the electron (usually called a majority carrier). The *bi* in *bipolar* refers to the device having both majority carriers and minority carriers. As already mentioned, the majority carriers are the electrons. The minority carriers are areas called holes, which can best be explained by an example. Suppose we have a piece of wood with 10 holes drilled in it, and we place nine ping pong balls in a row on the board, as shown in Figure 5.2(a). The nine balls take up specific spaces on the board, and there is one empty space, or a hole. If we move the first ball on the right one space to the right, we have the condition shown in Figure 5.2(b). The balls have moved in one direction (left to right), and the empty space has moved in the other direction (right to left). In a bipolar device, the electrons move in one direction, and the vacancy they produce when they move appears to move in the opposite direction. That is where the bipolar characteristic comes into play. We investigate this characteristic in greater detail later in the chapter. For now, it is sufficient to say that in the bipolar device there are actually two modes of moving current.

Another term that comes about when we are dealing with RF and microwave applications is *field effect device*. Whereas a bipolar device has two means of transferring energy through it, the field effect device is classified as a unipolar system. That means there is only one carrier of energy in that particular device. Unipolar devices have many applications in RF and microwave circuits since high frequencies take a very short period of time to cover one cycle. If you have two means of transferring energy from one point to another (electrons and holes), it takes much too long to respond to those high frequencies. If there is only one means of transferring the energy (electrons only), it takes much less time and

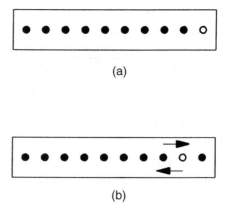

(a)

(b)

Figure 5.2 Examples of majority and minority carriers: (a) initial condition and (b) condition after one ball has been moved.

allows more than sufficient time for the device to respond to the applied frequency. Thus, the unipolar field effect device is used for many high-frequency applications.

We now present the two areas of RF and microwave solid state devices that find wide applications throughout the fields: microwave diodes and microwave transistors. We will look at the characteristics and operation of the Schottky diode, the PIN diode, the tunnel diode, and the Gunn diode. We also will describe bipolar and field effect transistors, as well as high electron mobility transistors.

5.1 Microwave diodes

When most people think of a diode, they think of a small, two-element device that is used to rectify an ac signal in order to get a dc voltage at the output of a power supply. That is basically what we described when we presented the diode vacuum tube in Figure 5.1. Such devices were not used for anything but rectification and occasionally for detecting a signal. For many low-frequency applications, that is still true. If you look at what diodes are available for audio and digital applications, you will see that these diodes are designated as rectifiers that pass current in one direction but not in the opposite direction. When you get into the RF and micro-wave spectrum, however, the situation changes drastically and the diode does much more than simply rectify an ac signal. This section shows that a simple two-element diode can amplify, oscillate, mix, detect, attenu-ate, and switch a high-frequency signal if used in the appropriate circuit. That may be hard to visualize if you are geared to low-frequency appli-cations, but by taking advantage of modern solid state technology, such devices do exist and are being used, even as you read this book. We will cover four types of diodes used in RF and microwave applications: Schottky (which mix or detect), PIN (which attenuate or switch), tunnel (which amplify or oscillate), and Gunn (which also oscillate). These high-frequency diodes are shown in Figure 5.3.

5.1.1 Schottky diodes

The Schottky diode gets it name from W. Schottky, who is known for his research in rectifiers in 1938. The Schottky diode has a different type of

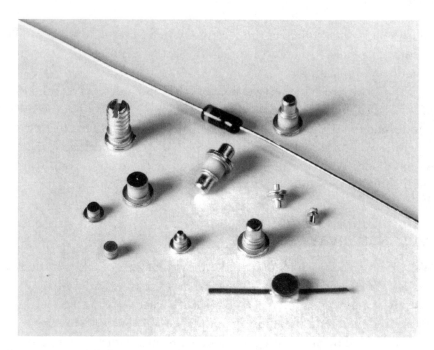

Figure 5.3   High frequency diodes.

construction from what usually is encountered with conventional diodes. The typical rectifying diode has what is called a *PN junction*, shown in Figure 5.4(a). In a PN junction, two semiconductor materials are "doped" with impurities to make one side primarily positive (P) and the other side negative (N). The idea of impurities and doping is very important to the operation of the PN structure. The base material for these devices is usually an "intrinsic" silicon. By intrinsic, we mean a pure silicon with no other material present. The intrinsic material then has certain impurities added to give it the properties that are either positive or negative by nature. In most cases, that is done by adding boron (B) or aluminum (Al) to make the material a P-doped material or by adding arsenic (As) or antimony (Sb) to make the material an N-doped material. When the doped materials come together, they form the junction shown in Figure 5.4(a). This device will now have bipolar properties with electron and hole movement as we mentioned earlier.

The Schottky junction, on the other hand, consists of a semiconductor (usually N-type) and a metal that come in contact with each other. This

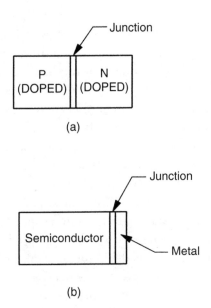

Figure 5.4 PN and Schottky junctions.

configuration can be seen in Figure 5.4(b). The metal used for these devices is usually aluminum, and the junction formed has unipolar properties, with only the majority carriers (electrons) moving the energy from one point to another. That is explained by considering that only one semiconductor comes in contact with the junction, so there is no hole movement across that junction. That is true because the material on the left in Figure 5.4(b) is a semiconductor, while the material on the right is a conductor (metal). That means the impurity factor described for the PN junction, which justifies the electron and hole movements, exists only on one side, that of the semiconductor. So only one type of conductor is present that controls the device, the majority carrier. That unipolar action makes the Schottky diode ideal for RF and microwave applications, because the diode can respond to high frequencies without distorting the output or causing other problems. It can respond faster because we do not have to wait for two carriers to perform before the device will operate. That means the time for the carrier to get across the junction, the transit time, is less than with the bipolar device.

The Schottky diode is the workhorse of the RF and microwave diode community. It is a general-purpose diode that usually is used for mixer and detector applications at those higher frequencies. Take another look at the diagrams in Chapter 4 of detectors and mixers, and you will see that each component has a diode or a diode circuit. Those diodes are Schottky diodes if they are operating at RF and microwave frequencies. This type of diode has many applications because it usually is low cost and readily available for a designer to use. It also is relatively simple to design this type of diode into a circuit. All these characteristics make the Schottky diode an excellent choice for many RF and microwave applications.

Figure 5.5 shows the equivalent circuit of a Schottky diode and the parameters that make up the circuit. These parameters are the series inductance, the series resistance, the junction capacitance, and a parameter called the overlay capacitance, which designated as C_o.

The series inductance, L_s, is the inductance of the bonding wires that go from the actual diode chip to the connections to the outside world. This parameter has typical values of 0.4 to 0.9 nH (nanohenries) (0.0000000004H to 0.0000000009H). This very small value is well controlled during the manufacturing process and usually is consistent from unit to unit.

The series resistance in the Schottky diode is the total resistance in the diode, including that of the semiconductor and the substrate on which the diode is mounted. The semiconductor is the material that comes in contact with the metal to form the Schottky junction. It has its own resistive properties that depend on the type of material. The substrate is the portion of the diode that is in contact with the semiconductor material

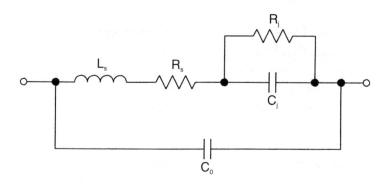

Figure 5.5 Schottky diode equivalent circuit.

on one side and the second terminal of the diode on the other end. It also has resistive properties that depend on the type of material used for the substrate. This physical relationship is why we say that the total series resistance of the diode is the total of the semiconductor resistance and the substrate resistance. Typical values for this parameter, R_s, are 4Ω to 6Ω. This parameter appears on Schottky diode data sheets. The value of the series resistance is particularly important when we are choosing a diode for a detector or mixer application. If the resistance is too high, power will be lost in the diode itself, and the maximum transfer of power cannot take place in the device.

The junction capacitance is the capacitance that is present across the actual junction between the semiconductor and the metal. It can be seen that with the Schottky junction a capacitance will be set up, since there are two plates with a dielectric material in between. The size of the junction determines how much junction capacitance actually will be in a device. Values for junction capacitance, C_j, range from 0.3 to 0.5 pF (picofarads).

The Schottky junction has another parameter that must be characterized and taken into account. That parameter is the junction resistance, R_j, the natural resistance of the area where the semiconductor and the metal come together. It stands to reason that some sort of resistance will be present at that point because two very different materials are suddenly forced together and allowed to have current flow through them. The junction resistance usually is a much smaller value than the series resistance of the diode and, as such, is not taken into account when we are determining the electrical properties of a particular diode. The main resistance is the series resistance, because it has a value that, if it were allowed to get too high, could hamper the operation of the diode.

The last parameter is the overlay capacitance, C_o, of the device. It is the value of capacitance produced from the Schottky junction to the metal contact of the opposite lead of the diode (not the lead where the Schottky junction is located). It is actually the capacitance across the combination of the semiconductor and the substrate (see Figure 5.6). It can be seen that this is an ideal capacitor in that it has two very distinct metal plates (the leads and the junction construction) separated by a well-defined dielectric between the plates.

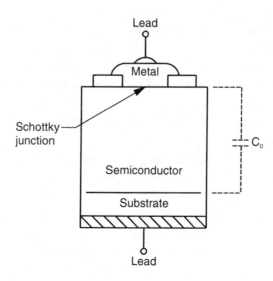

Figure 5.6   Schottky diode overlay capacitance.

As already stated, the Schottky diode finds many applications in detector and mixer circuits. If you recall from Chapter 4, a detector has the diode in series with the matching circuit and takes the input RF signal and detects it to allow a filter circuit to remove the RF and produce the required dc or video output. A mixer circuit has the RF and LO signals applied to the diodes that are used in the circuit. The diodes must be able to handle the LO signal level and also must be driven into nonlinear operation so mixing actually can take place. It is a good idea to have the diodes in the mixer matched as close as possible so the most efficient operation can take place. These diodes are good general-purpose diodes, but they also are very important parts of some important circuits, such as RF and microwave detectors and mixer circuits.

5.1.2 PIN diodes

Chapter 4 referred to the PIN diode in the description of the construction of an attenuator and a switch using PIN diodes and back-to-back quadrature hybrids. At that time, we simply said that when no bias voltage was applied to the diodes, there was a large resistance present, and when full bias was applied, there was a very low resistance for the device. We will now expand on those statements and explain the properties and operation of the PIN diode.

The PIN diode is actually a variation of the conventional PN junction. The main difference is a very small layer between the *P* and the *N* layers. This layer is called an intrinsic layer, which is where the *I* comes from in the PIN designation. *Intrinsic* can have a variety of meanings. In its truest sense, it refers to a perfect material, that is, a material with no impurities added. In semiconductor applications, silicon or germanium by itself would be considered a pure, or intrinsic, material. In actual practice, the intrinsic layer has some impurities, or doping, but substantially less than either the *P* or the *N* layer. That makes the intrinsic layer appear to be a perfect layer compared to the other layers. Because all materials have some impurities, it can be said that the designations *P*, *N*, and *I* are simply a matter of the total quantity of impurities in each material.

The PIN diode has another unique property. When the diode is operated at frequencies below about 100 MHz, it acts just like any other PN junction device and rectifies the signals that are applied to its input. Above that frequency, it acts like a microwave variable resistor. That is, as the bias on the device is changed, the resistance of the device is changed. Figure 5.7 is a plot of diode resistance as a function of bias voltage. You can see from the curve that with no bias applied a very high resistance is exhibited by the diode. As the bias is increased, the resistance decreases until, at some bias value, the resistance is very low. Thus, the diode looks

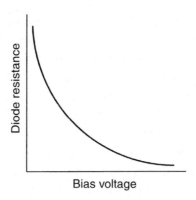

Figure 5.7 PIN diode curve.

like a high-frequency variable resistor and can be used as a variable attenuator for an RF or microwave circuit.

Some important terms associated with the PIN diode should be defined here, so you can see where the PIN diode might be used and when you have an acceptable PIN diode for the application at hand. These terms are *breakdown voltage, junction capacitance, series resistance,* and *carrier lifetime.*

Breakdown voltage, V_B, is a term that more or less defines itself. It is an RF voltage that is determined by the width of the I layer of the diode. This RF input voltage should not exceed the specification placed on the diode. If it does, the PIN diode's I layer will be punctured or destroyed, and the diode will then operate basically as a PN junction diode.

The junction capacitance is a little different from that same parameter of a Schottky diode. Junction capacitance for a PIN diode is designated $C_j(v)$. Notice that the capacitance is specified for a certain voltage condition; v indicates what voltage must be applied to result in the capacitance shown. A specification of $C_j(-50)$ gives you the junction capacitance that results when 50V is applied across the diode. The voltage must never exceed the breakdown voltage, V_B.

The series resistance of a PIN diode is the total resistance of the diode when a certain value of current is flowing through it. Just as the junction capacitance was measured at a certain voltage, the series resistance given in a data sheet is valid only if the resistance is checked at the specified current. The value of the series resistance, $R_s(i)$, usually is very close to being the minimum resistance of the device.

The parameter of carrier lifetime is shown on PIN diode data sheets as τ_L and is a measure of the ability of the diode to store an electrical charge. This quantity usually is expressed in nanoseconds and can be increased by eliminating some of the impurities in the device. The higher the concentration of impurities, the smaller the storage time, or carrier lifetime, for the device. If you are going to use the diode for a switch or an attenuator, you need to know how long the device will store a charge, so you can get some idea of what the response time will be for the diode to go from full off to full on. This is a critical parameter to know for proper operation of any component that uses PIN diodes.

The characteristics listed here make for some interesting applications of PIN diodes. Consider the circuit in Figure 5.8, probably the most

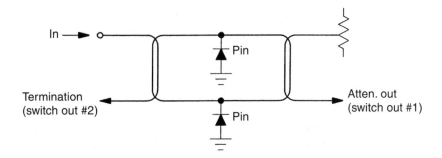

Figure 5.8 PIN attenuator/switch.

common application for PIN diodes. You will recognize this configuration from Chapter 4, in the discussion of quadrature hybrids. You will recall that in Chapter 4 this arrangement of back-to-back hybrids was called a constant impedance device. That is one of the reasons it is used here. Figure 5.8 shows back-to-back hybrids with two PIN diodes in between. The main difference between this figure and Figure 4.15 is that the circuit in Figure 4.15 has the PIN diodes in series with the quadrature hybrids, while Figure 5.8 has them in parallel, or shunt. The operation of the circuits is similar but necessarily must have a different bias-to-resistance relationship. In Figure 5.8, the operation is as follows. With no bias applied to the diodes, there is a very large resistance in the diodes, and basically the entire input signal appears at the output (the "Atten. Out" port in the figure). With some bias applied, there will be a smaller resistance, and some of the signal is now sent to the diode and not to the output. As the bias increases more and more, there is less and less signal at the output of the attenuator, which is exactly what we want. When there is full bias on the diodes, there will be very little signal at the output. We now have full attenuation of the attenuator. As a side point, the discussion of attenuators in Chapter 4 stated that for an attenuator to be efficient and useful it must attenuate and also maintain its VSWR over the full range of attenuation. The arrangement in Figure 5.8 does both those functions very well.

The circuit described here is a continuously variable attenuator, but it may be a step attenuator if the bias is applied in a specific manner to make the attenuation be in discrete steps.

Figure 5.8 also shows two other designations: Switch out #1 and Switch out #2. If we have no bias on the diodes, almost the entire input signal will appear at the Switch out #1 port. If we now go to a full bias condition, all of the input signal will be at the Switch out #2 port. Thus, we have created a switch with low insertion loss, high isolation between ports, and an excellent VSWR for both switched conditions.

So it can be seen that the PIN diode is not the everyday, ordinary PN diode that many people are used to. It accomplishes tasks that cannot be done with any other two-terminal device. Two functions are performed very nicely by the PIN diode: attenuation and switching. What distinguishes those functions from each other is how the bias is applied. It could be a continuous application of the bias to the diode(s), a series of steps, or an instantaneous change from zero bias to full bias. The method of bias application depends on what the particular application.

5.1.3 Tunnel diodes

The tunnel diode gets its name from the tunnel effect, which is a process whereby a particle, obeying all of the laws of quantum theory, virtually can disappear from one side of a potential barrier (basically a junction voltage) and appear instantaneously on the other side. This transfer occurs even though the particle does not appear to have enough energy to jump over the barrier. It is as though the particle can "tunnel" under the barrier. The electrons do get from one side of the barrier to the other and perform very well when they go through this process.

The tunnel diode barrier is the same as the junction region of a regular PN diode. The barrier in a tunnel diode is very thin, less than 1 microinch (0.000001 in). It is so thin that penetration by means of the tunnel effect is possible. The penetration results in additional current in the diode at very small forward bias. This very small voltage requirement is one that many designers use when power is limited, such in as portable transceivers and satellite applications, to name only two. The additional current in the device produces a negative resistance characteristic for the tunnel diode. That resistance is an ac resistance that is also called a dynamic resistance. It is important to understand that it is *not* a measurable dc resistance that can be measured with a meter and not a reverse indication on the meter. It must be calculated from measured values. This negative resistance appears because as the voltage applied to the diode increases past a certain point, the current ceases to increase and begins to decrease. That is in sharp contrast to the normal arrangement under Ohm's law that says that the current is directly proportional to the voltage applied. That is, an increase in voltage increases the current, and a decrease in voltage decreases the current (assuming the resistance of the device remains constant). Thus, the characteristic of negative resistance is an unusual one, but one that can be taken advantage of to allow the tunnel diode to perform some unusual functions that the normal PN junction cannot. Tunnel diodes are used for amplifiers and oscillators.

Two terms associated with the negative resistance of a tunnel diode are the minimum negative resistance, R_M, and the negative resistance at minimum K, R_N (K is the noise constant for the diode, also called shot noise). R_M is the smallest value of negative resistance exhibited by the diode. In Figure 5.9, it is the point on the negative side of the resistance curve (bottom portion) that comes the closest to either zero resistance or a positive resistance. Everything on either side of R_M is a greater negative value of resistance. The point on the current-voltage curve for a tunnel diode (Figure 5.10) shows where the voltage applied to the diode must be to obtain the minimum negative resistance. The minimum value of negative resistance is usually between 35Ω and 70Ω. It is interesting to

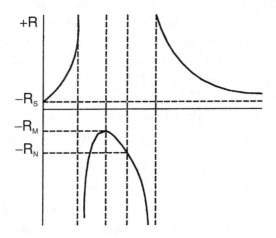

Figure 5.9 Resistance vs. voltage curve.

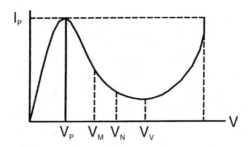

Figure 5.10   Current vs. voltage curve.

note that V_M is not the lowest voltage on the current-voltage curve, but it is on the downward slope of the curve. You can see that the minimum current in the tunnel diode creates an entirely different resistance.

The negative resistance at minimum K, R_N, is the resistance where the current is at its minimum value. This value depends on temperature, diode current, junction resistance, and the diode voltage at which the measurement is taken. Figure 5.9 shows where the value of resistance occurs. You can see that it is not the minimum negative resistance but some value that is down the curve away from the minimum value. The current-voltage curve in Figure 5.10 shows that the negative resistance at minimum K occurs when the current reaches its minimum value. Thus, the term actually gives the value of negative resistance for the best noise performance of the device, which happens when the current is at its lowest point. That makes sense, since the lower the current in any device, the lower the noise. R_N is a higher value of negative resistance than R_M; values can range from as low as 40Ω to as high as 120Ω.

The value R_s in Figure 5.9 is the series resistance of the tunnel diode. This is a constant value that is lower than any value of resistance shown on the main curves. It is also a positive resistance and depends on the structure of the diode.

Typical applications for a tunnel diode are for tunnel diode amplifiers, tunnel diode oscillators, and tunnel diode detectors. The most common use of the diode is for an amplifier. Recall that Figure 4.44 showed a block diagram for a tunnel diode amplifier when we were covering circulators and isolators. The amplifier and circulator only are reproduced here in Figure 5.11. It can be seen from the figure that the input signal is applied

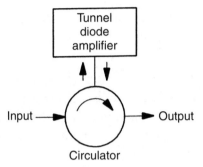

Figure 5.11 Tunnel diode amplifer.

to the input of the circulator. Because of the circulator motion, which follows the arrow on the schematic diagram, the signal is sent to the tunnel diode amplifier circuit. With the negative resistance characteristics of the tunnel diode, the signal sees a large mismatch and is reflected back into the circulator, where it is sent to the output port. The reflections and the circulator operation produce an amplification.

The main advantage of a tunnel diode amplifier is that it operates with very low voltage. It is not unusual to have a voltage requirement of only 125 mV, as opposed to +12 or +24V for other types of amplifiers. The amplifiers deliver excellent gain, and the tunnel diode amplifier operates at frequencies up to 20 GHz. The main reason that many of these devices were replaced was that their main use was for satellite applications before field effect transistors were introduced. The tunnel diode's low power was ideal for the inefficient solar panels used at that time. However, as the solar panels became more and more efficient, field effect transistors found more applications since the power was available for them. Another drawback for satellite applications soon became apparent. The circulators that were used were very large in some cases, which was not what satellite designers wanted since the advantage of low power was now gone. Even though not many satellite applications still exist for tunnel diode ampli-fiers, there are still some ground-based systems for which these amplifiers are ideal.

The tunnel diode is one device that is still constructed on germanium. If high gain and moderate noise characteristics are the goal, the germa-nium device should be the choice. If the circuit is one that must have some very low noise characteristics, gallium arsenide (GaAs) is the material that should be used for the tunnel diode.

5.1.4 Gunn diodes

In the 1980s it was said that as long as the properties of semiconductors depended on junctions and those junctions had to be made thinner as the frequencies increased, high-power semiconductor devices at microwave frequencies never would be possible.

Well, the adage "Never say never" applies here. Devices have been developed that do not depend on a junction for operation. One such device is the Gunn diode, which is in a class of devices that exhibit microwave power properties that depend on the behavior of bulk semiconductors rather than junctions. Those semiconductors, for our applications, are GaAs and indium phosphide (InP). The Gunn effect is the main representative of that group of devices, which was discovered in 1963 by J. B. Gunn. He found that when a dc voltage is applied to the contacts on the end of N-type GaAs or InP, the current first rises in a straight line (linearly) from zero and then begins to oscillate when a certain threshold is reached. The time of those oscillations, that is, the period for one cycle, is very close to the time it takes for carriers to travel from one contact to another through the N-type material. This effect became known as the *Gunn effect*, or *bulk effect*. The Gunn diode is the main device that utilizes this effect to its fullest.

Calling the device a diode, however, is confusing, because there is no junction, which all other diodes exhibit. Figure 5.12 is a cross-section of a Gunn diode, showing the individual layers. It can be seen that there are only N layers present in this device. (The plus and minus signs indicate how heavily the N-type material is doped; a plus sign means it is more heavily doped, a minus sign means less doping.) There is no PN junction to be seen anywhere. Actually, the device is called a diode simply because it has two leads, just like all other diodes. Also, it is a more common way of referring to the device rather than simply calling it a Gunn device. The term diode has a much better ring to it and is more recognizable.

The Gunn diode, like the tunnel diode, is a negative resistance device. Recall that this property shows up in the voltage-current relationship. When the bias voltage is applied to the diode, it initially causes an increase

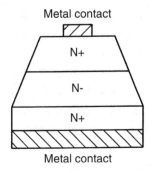

Figure 5.12 Gunn diode.

in current through the device. As the bias voltage is increased more and more, the diode reaches a point where a further increase in bias will cause the current through it to decrease. That relationship is shown in Figure 5.13, which is an I-V curve for a Gunn diode. It can be seen from the figure that as the bias voltage is increased the resulting current also increases for the first part of the curve. This is a nice linear curve that behaves very well. As we get to point A, the bias voltage is still increasing, but you will note that the current now is decreasing. Point A is where the Gunn effect kicks in and causes the device to become the negative resistance device that it is. Once the current starts to decrease, there is a very straight portion of the curve from point A to point B. That is where the diode should be operated. That nice linear area, even though it is a

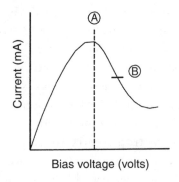

Figure 5.13 Gunn diode I-V curve.

negative resistance area, ensures that you are operating in the optimum region for the diode to be the most efficient and able to generate a stable frequency and power output level.

The primary application for a Gunn diode is as an oscillator. Many times the LO for a microwave receiver will be a circuit with a Gunn diode. This is a primary communications application. One area where the Gunn oscillator also is used is for sweep generators to test RF and microwave circuits.

Applications for Gunn diodes extend beyond the microwave spectrum, well into the millimeter range of frequencies. Gunn diodes also can be used for radars, beacons, transponders, speed sensors, radio, and data links. To cover that much range, the material used for construction of the device must be considered. For a microwave oscillator, the diode should be constructed of GaAs. For higher frequencies, in the millimeter frequency range (above 18 to 20 GHz), the diode should be constructed of InP. Gunn diode oscillators provide an excellent and stable output for the circuits or systems they must feed. They also exhibit some very good noise characteristics.

5.2 Microwave transistors

The microwave transistor is the device that probably has advanced and improved the most of any microwave device. Even though many new diodes have been introduced, the transistor has advanced and improved on its basic designs and structure while introducing basically only two new types. This section describes those two devices, along with the conventional bipolar device that started the whole process.

The first transistors were fabricated by Drs. W. H. Brattain and J. Bardeen in 1948. The discovery occurred while they were studying the properties of germanium semiconductor rectifiers at Bell Laboratories. During their studies, Brattain and Bardeen observed that the flow of current through a rectifier could be controlled if a third electrode was added to the device. That electrode served much the same purpose that the control grid did for the triode vacuum tube. Until these experiments were conducted, the only devices available that would rectify a signal were two terminal germanium devices. That was equivalent to the diode vacuum tube. The vacuum tubes needed to be controlled, as did the solid state devices that were being looked at in the 1940s.

The diode was improved by placing a control grid between the cathode and the anode of the device. By placing various voltages on the grid, it was possible to control the flow of current in the tube. In the same manner, the experiments that were to lead to the first transistors were designed to have a third element added that would control the flow of current through the germanium crystal. What resulted from those experiments was the point contact transistor (see Figure 5.14). This device was a major breakthrough for the electronics industry because it promised to replace many of the bulky, hot, and high-power-consuming vacuum tubes of the day.

The point contact transistor, however, had drawbacks. This revolutionary device generated much more noise internally than the vacuum tube it was designed to replace. Also, it was not hermetically sealed and thus could not tolerate temperature or humidity and was very fragile to handle. In 1949, those problems were addressed when W. Schockley published a paper with a description of the junction transistor, the basis for most of the transistors in use to this day. Figure 5.15 shows the junction transistor and its different type of structure. You can see that there are much stronger contacts for the leads of the base collector and emitter because they are connected through a metal contact as opposed to being directly connected to the P material, as the point contact transistor was. The junction transistor has a much improved noise

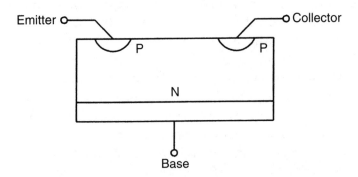

Figure 5.14   Point contact transistor.

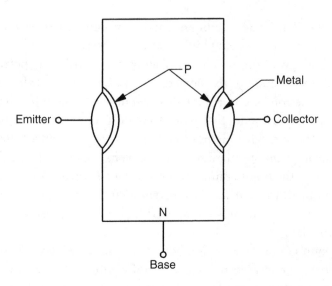

Figure 5.15 Junction transistor.

performance and was considerably more rugged than the point contact transistor.

To clarify some of the terms used in reference to basic transistor constructions, we should briefly explain the contacts shown in Figures 5.14 and 5.15. The input to a transistor (bipolar for now) is the base lead. This is the control element in the transistor where the ac signal or, possibly, a dc voltage is applied to control operations in the device. The emitter is the source of electrons to the transistor and can be related to the cathode in a vacuum tube. The collector is the output lead of a transistor and can be related to the anode of the vacuum tube. These explanations are for what is called a common-emitter type of configuration, which has the emitter going to ground either directly or through a resistor. There are other arrangements in which the emitter is the output, but those are special cases and are touched on only briefly in this text.

The types of microwave transistors that evolved from the junction transistor are *planar* and *epitaxial* transistors, the most common devices in use today. The planar device, as its name implies, has all its elements in the same plane. Figure 5.16 is a basic diagram of a planar transistor. The doped materials are diffused into the device at the proper locations to form the base and emitter sections. (Diffusion is the combining of

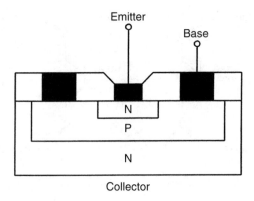

Figure 5.16 Planar transistor.

materials by forcing one into the other.) The collector element is at the bottom of the device. Notice how everything is lined up, or planar, in this device.

The epitaxial device is similar to the planar transistor, with the exception of an additional thin layer of low-conductivity material in the collector region. In Figure 5.16, that layer would be placed between the N layer of the collector and the P region used for the base regions. The collector region, as it appeared in the planar package, is now a high-conductivity area designated as N^+, with the epitaxial layer (low-conductivity area) designated simply as N. This type of construction sometimes is referred to as an epitaxial-collector type of device, which often is its designation on data sheets. The epitaxial device is probably the most widely used form of high-frequency transistor construction used today.

We now will look at three types of microwave transistors in common usage today: bipolar transistors, field effect transistors, and high electron mobility transistors. Each type is presented and discussed to help you to become familiar with how the devices do their jobs and how they can be used for particular applications.

5.2.1 Bipolar transistors

Figure 5.17 shows two schematic representations of bipolar transistors, one of an NPN device, the other of a PNP device. The way to distinguish

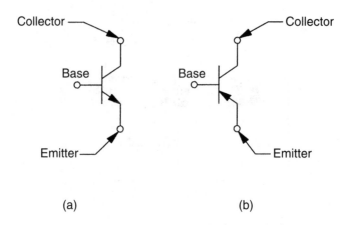

Figure 5.17   Bipolar transistors: (a) NPN and (b) PNP.

between the devices is to look at the arrows. For the NPN device, the arrows point out of the device. A quick way to remember that is to say, "Not Pointing iN." The PNP device, on the other hand, has the arrows pointing in. You might remember that by thinking of the arrow as Pointing iN toward the device (and just sort of ignore the final P). The voltage applied to the output terminal (collector) is positive for the NPN device and negative for the PNP device. An easy way to remember which is which is to use the middle letter: P in the middle of NPN indicates a positive collector voltage, and N in the middle of PNP calls for a negative voltage on the collector.

Figure 5.17 also shows the terminal names for the bipolar transistors. These were covered previously but are briefly explained again for clarity. The input terminal is the base of the transistor, the output terminal is the collector, and the third terminal is the emitter (the origin of the electrons that are sent to the collector). In most applications, the signal to the transistor is applied at the base of the device, with the emitter either grounded or having a resistor bypassed with a capacitor. The output of the device is taken from the collector, which is connected through a resistor to the collector supply (positive or negative, depending on the type of transistor). This configuration is called a common emitter.

Figure 5.18 shows a typical common emitter circuit. The additional resistors (R_1 and R_2) shown in Figure 5.18 are for dc biasing the transistor to allow it to operate in the proper region of its characteristic curves. This

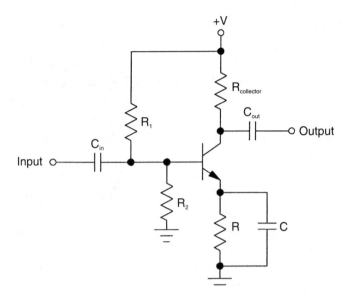

Figure 5.18 Typical bipolar circuit.

is called a voltage divider biasing network and is used for many low- and high-frequency amplifier circuits. The input capacitor (C_{in}) and the output capacitor (C_{out}) are in the circuit and are called coupling capacitors. Their function is to couple the high-frequency signal into and out of the device and not allow the dc voltages to interfere with the preceding or proceeding circuits. This is an excellent circuit that is easy to build and produces a clean, amplified signal at the output if the biasing is set properly and the coupling capacitors are the proper value.

One type of bipolar transistor used in many RF and microwave applications is the low-noise transistor. This type of device must supply a certain amount of gain for a circuit and, at the same time, exhibit a very low level of noise. The way that the low-noise transistor provides a low-noise characteristic is through the construction of the device. The emitter and base regions of the transistor are fabricated in an interdigital construction. Chapter 4 described that type of construction in the discussion of bandpass filters. Figure 4.34 presented three types of filter constructions, the third one of which was a series of resonators a quarter-wavelength long and grounded on alternate ends. That last characteristic, the grounding on alternate ends, is what makes the filter an interdigital filter. A similar construction is used for low-noise transistors and is shown in Figure 5.19. The emitter and the base of the transistor are split into a series of "fingers." That splitting of the paths for current is helpful in getting low-noise characteristics, because the best noise source you can get is a resistance with current flowing through it. It is important to note here that we use the term *resistance*, not *resistor*, when we talk about generating noise. It is true that current through a resistor generates noise, but that is a very special case. The more general—and more realistic—case is when you have any type of resistance in the circuit through which current is flowing. That produces a noise that must be considered. Thus, if we can reduce the resistance of a device or decrease the current flowing through it, we can reduce the noise and create a low-noise device, such as a transistor.

The emitter-base junction of a transistor is a resistance with current flowing through it. If that resistance can be lowered, or the current flowing through it can be reduced, the overall noise in the transistor is lowered. That is exactly what the inderdigital structure shown in Figure 5.19 does. It can be seen in the figure that each emitter portion of the transistor is divided into separate sections, with each section having specific resistance. Because all the fingers are parallel, the total emitter resistance is much smaller than any one of them individually. If all the fingers are the same resistance, the total resistance is the resistance of one of them divided by the total number of sections in the transistor. That greatly decreases the total resistance of the emitter-base region and is an excellent start in creating a very low noise device. Figure 5.20 shows the internal structure of a bipolar transistor.

In the same sense, the total current at the input to the interdigital structure is divided among all the sections of the structure. That results

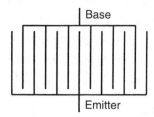

Figure 5.19 Interdigital construction.

Figure 5.20 Internal structure of a bipolar transistor.

in each individual section carrying much less current than if a single section for the junction is used. With both of those parameters, resistance and current, reduced by use of the interdigital structure, interdigital construction significantly reduces the noise characteristics RF and microwave transistors.

The low-noise transistor has many applications in communications receivers. Of the applications listed and discussed in Chapter 2 for wireless systems, many areas require a low-noise receiver to improve performance and increase the coverage range of the system. Thus, a low-noise device must be right up front to guarantee that there are no problems with noise in the system. Often you might see a combination of two low-noise transistor stages at the receiver input with a quadrature hybrid (or Lange coupler, which is the microstrip version of the quadrature hybrid) used to supply the input signal to both of the devices, as shown in Figure 5.21. So, you can see that the low-noise transistor finds many critical applications in RF, microwave, and wireless systems.

Another common bipolar RF and microwave transistor is the power transistor. The power transistor produces the high-power output levels for final amplifiers of RF and microwave transmitters. These transistors are much different from low-noise transistors, primarily because they must carry a much higher current than the low-noise devices. Recall from our discussions of low-noise transistors that we worked hard to obtain a low current through a low resistance to get low-noise characteristics. In the power transistor, there must be a high current to produce the required

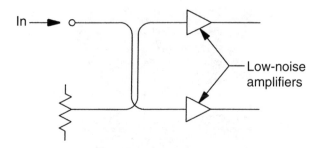

Figure 5.21 Low-noise input.

high output power. That is because power equals voltage times current. The voltage for the power transistor is somewhat higher (e.g., collector voltages of +24V, as opposed to +10 or +12V for a low-noise device), but the voltage is not the primary parameter used to produce power in a device. That distinction goes to the current. The higher the current, the higher the power.

Because we need to handle more current in a power device (1A to 2A, compared to a few milliamps for a low-noise device), the power transistor needs to be much more rugged and requires much larger elements than its low-noise counterpart. You will notice that right away when you look at a low-noise transistor and a power transistor side by side (see Figure 5.22). The first device is a low-noise transistor and shows a small area for the device with some narrow leads for the base, the emitter, and the collector. This type of device usually is put in the common emitter configuration, which has the input on the base and the output on the collector. The second device is the power transistor and has a different configuration. These devices usually are built as common-base circuits, an arrangement in which the input is at the emitter and the output at the collector. The "common" element is the base. That can be seen in Figure 5.22 as that portion of the device connected to a flange and attached to a base plate by two screws (designated by the two circles, which are through holes) in the base leads. The common-base configuration is used so the necessary powers can be generated for a particular application.

Note in Figure 5.22 that the power transistor has leads on it that are much wider than in the low-noise device. Also, the package itself is much

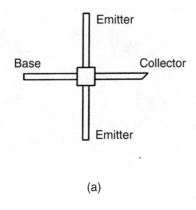

(a)

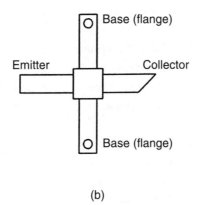

(b)

Figure 5.22 Low-noise and power packages: (a) Low-noise device and (b) high-power device.

larger and with a flange attached to it. The flange not only is an electrical connection but is also a connection to remove the heat from the transistor itself. Used as a heat sink, the flange gets any dissipated power inside the device away from the transistor junction and is a good connection to the much larger heat sink that is used for the entire power amplifier. A source for heat removal from a power transistor is an absolute necessity for improved reliability. The more heat that can be removed, the longer the device will last. Figure 5.23 shows the internal construction of a bipolar power transistor. Notice the size of the elements compared to those of the low-noise device shown in Figure 5.20.

Figure 5.23 Internal structure of a power transistor.

Another bipolar transistor, the linear transistor, is one that sort of falls between the low-noise, low-current device and the high-power, high-current power transistor. The linear transistor is designed to amplify signals that come from the low-level transistor stages and go the high-power output stages of many systems. It is called a linear transistor because it operates in a very linear region of the power output versus power input curve (Figure 5.24). The linear curve was presented in Chapter 4 in the discussion of mixer circuits. At that time, we were trying to get a circuit to operate in a nonlinear area of the curve and ignored the linear portion. Now we are looking specifically at the linear region of the curve, which is a straight line that is very predictable and well-behaved.

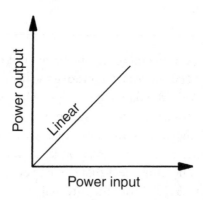

Figure 5.24 Linear curve.

That is what we looked for when we were working with the low-level, low-noise devices to have a "clean" output signal from the amplifier, that is, an output that was an amplified version of the input signal with no distortion. This is the same thing we are looking for with the linear transistor, except that the output levels are much higher than those with the low-noise transistors. These are the missing blocks that get the low-level signals high enough to drive the power stages, which produce an output power only when the input is a sufficient level to turn them on.

One difference that can be noted on transistor data sheets for linear devices is the operating current level. For low-noise devices, the primary consideration is to keep the current low so the noise level does not get too high. For linear devices, noise is not the primary parameter to be considered; what needs to be obtained is more gain. With that in mind, we can check the data sheets for a low-noise device and for a linear device and find that the low-noise device operates with currents of 5 to 10 mA, while the linear device probably is in the 200- to 300-mA region, which certainly is not a low-noise condition. But, as we have explained, the linear device usually is not a low-noise device. It is, rather, a high-gain device designed to produce the intermediate power levels that drive power amplifier circuits.

If you look at a linear transistor, you will see that its size and the size of its leads fall between those of a low-noise transistor and a power transistor. The leads must be large enough to handle the increased currents of the low-noise transistor, but they do not need to be as large as the power transistor leads shown in Figure 5.22.

Linear transistors are devices that truly fall between the low-noise and power devices we have discussed. They find many applications in RF and microwave systems where a low-level signal must be amplified to drive a power amplifier to achieve a specified power output.

For many years, bipolar transistors were the only transistors available to designers, and they still find many applications in RF and microwave systems and circuits. They are not, however, the only game in town anymore, and that is how it should be. Other forms of transistors used for these applications will now be presented. The idea is to evaluate the properties of bipolar devices and other transistors and decide which is the best for a particular application.

5.2.2 Field effect transistors (FETs)

The *field effect transistor* (FET) came into its own in the 1970s and was a buzz word for many applications during that time. The theory for the FET, however, goes back to 1926. In 1952, William Schockley first proposed them as devices that could solve a number of problems. At that time, however, many technological and fabrication difficulties kept the FET from arriving on the scene until around the 1960s. By then, however, the silicon bipolar device was so well defined that it pushed FET development further into the 1970s. The FET has had a rough history before the need and the technology got together to produce one of the best semiconductor devices since the first germanium diode was constructed.

One of the earliest FETs became available about the same time as the bipolar was introduced. That transistor was the *junction* FET (JFET). This device was a low-frequency transistor and did not really challenge the bipolar transistor at the time of its introduction. Many advances in techniques for designing and fabricating FETs led to the development of the *metal-oxide-semiconductor* FET (MOSFET). This device also is referred to as an *insulated-gate* FET, or IGFET.

The FET used for RF and microwave circuits is the *metal-semiconductor* (MESFET). This device uses a Schottky junction (metal-to-semiconductor junction). An end view of a MESFET is shown in Figure 5.25. A much different arrangement of the elements makes up this type of transistor than for the bipolar devices. First of all, the designations for a FET are different from those of a bipolar device. Recall that a bipolar transistor has a base, an emitter, and a collector. The FET uses a gate (base), a source

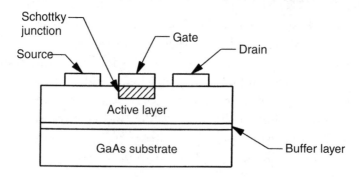

Figure 5.25   MESFET.

(emitter), and a drain (collector) for its terminals. The source, as the name implies, is the source of electrons, which, for the bipolar device, was the emitter of electrons; the drain is where the electrons go to, similar to the bipolar's collector; and the gate controls the current flow from one element to the other, just as the base did for the bipolar. One way to remember the designations for the FET is to look at the device and say that the electrons come from a source, a gate is used to control the flow of current, and the electrons must go down the drain to escape.

Figure 5.26 shows two types of FETs: an N-channel and a P-channel. The N-channel FET is similar to an NPN bipolar transistor in that it takes a positive drain voltage. Similarly, the P-channel FET is like the PNP transistor, since it uses a negative drain voltage.

Referring back to Figure 5.25, we can see that the layout of the elements on the FET is different from the layout of a bipolar transistor. A bipolar device has the emitter and base terminals on the top portion of the device with the collector at the bottom of the device. A FET, on the other hand, has all its elements—the source, the gate, and the drain—on the top of the device. Also, the FET uses the Schottky junction, which can be seen directly beneath the gate element. The area under the FET elements, called the active layer, is the semiconductor area and makes contact with the metal element of the gate to form the Schottky junction. The FET is completed by the addition of a buffer layer of semiconductor material and placement of the entire assembly on top of a GaAs substrate.

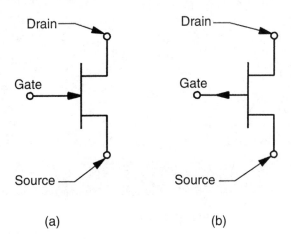

(a) (b)

Figure 5.26 FET schematics: (a) N-channel and (b) P-channel.

The FET, like the bipolar, comes in a low-noise version. The construction of the low-noise device also emphasizes a low resistance and a low current through the transistor. Because the geometry of a FET is different from that of a bipolar, the interdigital structure is not used. Instead, the device keeps the gate and the emitter well proportioned in regard to each other, and as the structure appears it exhibits low resistance and current. Figure 5.27 shows the internal structure of a low-noise GaAs FET.

A couple of terms must be understood when we are dealing with low-noise FETs. The first term is, of course, *noise figure*. On a data sheet, that may be termed optimum noise figure (NF_{OPT}), minimum noise figure (NF_{MIN}), or spot noise figure. Each of those terms refers to the noise figure when the conditions listed, drain-to-source voltage (V_{DS}) and drain current (I_{DS}), are noted and complied with. The noise figure values do not occur if the conditions are not duplicated as they were during the initial tests of the transistor.

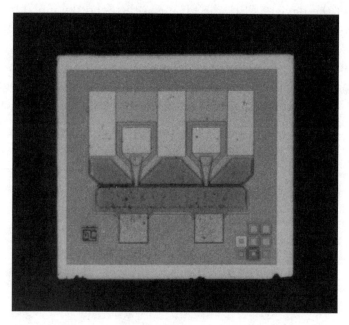

Figure 5.27   Internal structure of a low-noise GaAs FET.

Another term that is of great value when looking at FETs is *gain*. Usually, two gain numbers are involved when we are talking about, low noise FETs: maximum available gain (G_{MAX} or *MAG*) and gain at optimum noise figure (G_{NF}). Actually, the maximum available gain is not considered in a low-noise application. It does come into play, however, in a low-level application that needs a certain amount of gain. The current for this value of gain is higher than that for the low-noise applications.

The gain at optimum noise figure is a gain that occurs when the specified noise figure is reached. That gain, as you might reason, is less than the maximum gain because less current is being drawn, so a low-noise condition occurs. This gain must be known if a low-noise circuit is the goal.

Power FETs, as might be expected, are much different structures than low-level or low-noise devices. The power FET has many areas on the chip so that each element can carry its share of current. That is similar to the bipolar power transistor structure, which also has many areas to carry the current. The main purpose of a power FET is to produce a high-power output, which is accomplished by producing and carrying a high amount of power. Figure 5.28 shows the internal structure of a GaAs FET power transistor.

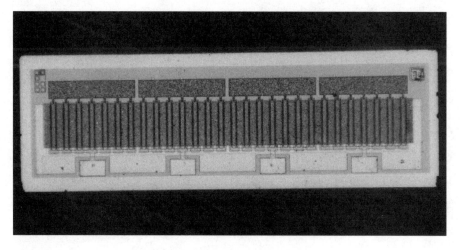

Figure 5.28   Internal structure of a power GaAs FET.

Three terms apply to FET power transistors. The first term is the *output power*, which is presented in two ways: as the straight power output of the device and as the power output at what is called the 1-dB compression point. The straight power output tells what the linear power is, that is, the power that is at the output with a specific input power. As the input power is increased, the output power follows it. That is the linear power output and is a parameter that has been discussed many times throughout this text. The familiar P_{out} versus P_{in} curve is repeated here as Figure 5.29.

Figure 5.29 also describes the second type or power output specification that appears on a GaAs FET power transistor data sheet, the power output at the 1-dB compression point. The 1-dB compression point is shown in Figure 5.29 as the point where the curve starts to level off. It gets its name from the fact that as you apply an input to a device you get a certain output that is dependent on the gain of that particular device. If you keep increasing the input level by 10 dB, you get a corresponding increase in the output. There is a point where a 10-dB increase in the input power produces only a 9-dB change in the output. That point is the 1-dB compression point. The 1-dB compression point also is the beginning of the nonlinear portion of the power input versus power output curve.

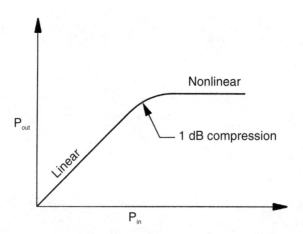

Figure 5.29   P_IN vs. P_OUT curve.

The gain of a power transistor is an important parameter to consider. It would be rather crazy to supply dc power to a power transistor, work hard to get an appropriate input power level for that device, and then have the transistor have a gain of only 1 or 2 dB. There must, therefore, be a sufficient gain either to produce the required output power for the system or to get very close to it so that an additional stage will accomplish the task. Most GaAs FET power transistors produce at least 7 to 10 dB of gain for the frequency ranges used for RF and wireless applications. That usually is sufficient to accomplish the tasks that need to be done for these RF systems. Be sure you know the gain of the device you will be using and always check to see at what frequency the gain is being specified. If you are going to need 10 dB of gain at 1.9 GHz, for example, it does you no good to have a transistor that has a maximum frequency of operation of only 1 GHz but has 15 dB of gain there. Be sure to check the gain and the frequency for that gain.

When we are dealing with power transistors, we need to consider parameters that do not apply to low-level or low-noise devices, that is, the thermal properties of the device. With high-power transistors, there are much higher drain currents and, thus, an elevated temperature within the transistor itself. One of the primary parameters is the thermal resistance, expressed in degrees Celsius per watt, which tells you how well the device gets the heat generated inside the device out of the device. A power device has a chip, which is the actual device, attached to a substrate, which is attached to some sort of flange, which attaches to a chassis. The heat must go from the chip to the chassis in the shortest amount of time and encounter the least resistance possible. That is where thermal resistance comes into play. It is the amount of resistance that the thermal energy will encounter as it makes its journey to the chassis. The value, which should be as small as possible, depends on temperature and power, thus the measurement in degrees Celsius per watt. So consider how much the temperature will be elevated with the power you will be generating when you are choosing a device to be used for a power application.

The FET is a device that was designed to solve many of the problems that the bipolar transistor had for some applications. It actually did so, but that is not to say the FET has completely replaced the bipolar for every application. There is room in the RF and microwave world for both devices.

5.2.3 High electron mobility transistors (HEMTs)

The latest addition to the lineup of solid state devices is the *high electron mobility transistor* (HEMT). This technology has come through the typical growing pains that both the bipolar and the FET had to encounter as they were being developed. When HEMTs first were proposed in 1978, they received an indifferent reception. Since then, they have attained a much higher credibility in the industry simply by demonstrating their performance in actual applications. In some areas, they have outperformed the GaAs FET, while in other areas they have not. So, it is the typical situation in which no one device is good for all applications. The HEMT is available both as a low-noise device and as a power device. It was not originally developed to compete with the GaAs FET power transistors, but it has found many applications where it does just that.

The HEMT looks similar to the low-noise FET. Their gate-source-drain structures are almost identical. So, what is it that makes this device any better than the conventional FET? The answer is that the difference is not in the structure but in the semiconductor materials. HEMTs use a doped GaAs/AlGaAs structure in which the motion of the charge carriers is confined to a thin sheet within the GaAs buffer layer. The type of structure results in the electrons within the transistor having a significantly higher degree of freedom. The freedom results in a higher mobility for the electrons (which is where the name comes from) and having them able to respond much faster for use at higher RF and microwave frequencies.

HEMTs have many applications in RF and microwaves. Their claim to fame is the lower noise characteristics at higher frequencies. For example, a GaAs FET may have a noise figure of 1.0 dB at 5 GHz, but the HEMT will have a noise figure in the range of 0.4 to 0.5 dB. The gain of the HEMT, however, will be less than that of the FET. At that same 5 GHz, the FET will have a gain in the neighborhood of 20 dB, while the HEMT will have around 15 dB. These values can be seen in Figure 5.30. Looking at the curves in the figure, you can see that there are tradeoffs in the decision of which device to use for which application. Some applications require the low-noise figure, while others do not need that as a critical parameter. On the other hand, some applications need as much gain as possible. The choice of HEMT or FET depends entirely on the application.

As with other solid state devices, certain parameters on the data sheet must be taken into account, for instance, the minimum noise figure. That parameter is, as the name implies, the minimum-value-of-noise figure for the device. This parameter does not appear by itself; it always is accompanied by the conditions it was taken under (V_{DS} and I_{DS}), as well as the frequency at which the data were taken. Usually, a variety of frequencies are used and, as a result, a variety of noise figures. It can be seen from Figure 5.30 that the noise figure is fairly flat over the frequency range shown, but there are some variations from one end of the range to the other.

Gain is also an important parameter for HEMT. Associated gain figures and maximum available gain figures usually are on HEMT data sheet. The associated gain is the gain that results when the value-of-noise figure listed in the minimum noise figure category is achieved. This is the same parameter that was described for other devices. The maximum available gain is just that, the maximum gain that is available from the

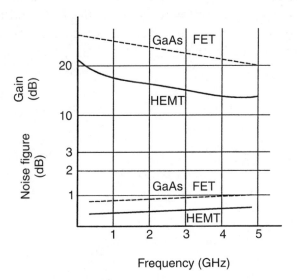

Figure 5.30 GaAs FET vs. HEMT.

device if the noise figure is not a consideration. There may be a difference of 5 to 6 dB in these figures. There also is a much greater difference in the noise characteristics when you consider the gain only of the device you are using.

Thus, the HEMT joins its predecessors in the RF and microwave field. This device has had—and continues to have—its ups and downs, its triumphs and defeats, just like the bipolar and the GaAs FET. The HEMT, just like the bipolar and GaAs FET, has found its place in the technology and will find many more applications in the future.

5.3 Summary

This chapter presented the world of RF and microwave from the perspective of high-frequency semiconductors. A general introduction presented some critical definitions and standards regarding RF and microwave semiconductors. Microwave diodes include the Schottky, PIN, tunnel, and Gunn diodes. The area of high frequencies is the only area where a two-element device can amplify, oscillate, attenuate, switch, mix, and detect.

This chapter also covered microwave transistor: the bipolar transistor, the first type of transistor to be used; the FET; and the HEMT. Designers have a great many solid state devices to use in today's sophisticated designs and applications.

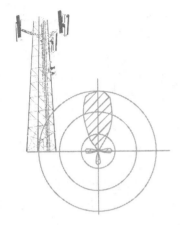

6

Microwave Materials

ICROWAVE MATERIALS have advanced to a new level over the past 10 to 15 years. Many years ago, only one material was available to RF and microwave designers. That material was a woven Teflon fiberglass material that came in only one thickss, haharacand was used for every application, regardless of the frequency ranges, power levels, or loss requirements that were called out. Today, there are many types of materials with a variety of parameters, so materials can be custom fitted to just about any application. With a variety of materials available, there is no excuse for using the same material for every application. No one should rely on a pet material for all designs. The mind-set "We've always used this material, why not use it again?" or "This material worked great for our last job, so we might as well use it again; why take chances?" does not work today. There is no excuse for not having a working circuit, provided that the right design procedures are used, if the time is spent choosing the proper material for the application at hand.

Microwave materials are special types of materials. These materials are an integral part of the circuits they support, not like the low-frequency and digital circuit boards (those with low data rates), where the only functions are to support components and get current from one point to another. The microwave material is actually a part of the circuit that is being designed. This was shown briefly in Chapter 1, when we distinguished between lumped and distributed components. Also, recall from Chapter 3 that microstrip and stripline transmission lines are constructed on a circuit board material and that the required lengths and widths of the transmission lines are determined by the characteristics of the material on which they are placed. So you can see that microwave circuit boards are not ordinary in any respect and that they do much more than support components. The circuit board material is an integral part of the RF and microwave circuits for which they are the base. A common thought in the field of RF and microwaves that designers should spend almost as much time choosing a material to use as they do to design the final circuit. That is how important the material is in these high-frequency circuits.

6.1 Definition of terms

As in any topic, the terminology used to describe microwave materials is important. Specific terms need to be defined and understood before any meaningful discussions can take place. The terms that we look at in this section are *dielectric*, *dielectric constant*, *dissipation factor*, *dielectric thickness*, *peel strength*, *copper weight*, and *anisotropy*. All these terms appear on material data sheets and should be understood by anyone involved with microwave materials. Some alternative terms may be used by engineers; these are the same terms, just expressed differently than some of the conventional terms.

The first term is *dielectric*. The dictionary describes it as an insulator, that is, a device or material that does not allow current to pass through it. That is an accurate description of a dielectric as it applies to power line insulators or the material in capacitors. It is not, however, an accurate definition for use in microwave materials. To understand the difference, consider the following example. A large pipe is suspended by wires with a target attached to the wall at the end of the pipe. If the pipe is empty, we can throw a ball through it and hit the target fairly easily. If we fill the pipe with some loosely packed feathers and throw the ball toward the target, more energy is required to get the ball to the target and it will get there a little slower than when the pipe was empty, but it probably will get there. If we fill the pipe with water (and use a bit of license to pretend that the water does not spill out of the pipe at both ends), it will be very difficult to get the ball to the target.

In our example, the ball did not change, the pipe did not change, the target did not change. The only thing that changed was the medium that we had to throw the ball through. In the second and third case, we slowed the ball and impeded its path to the target. We *obstructed* its path. That is what a dielectric does when it is applied to microwaves. It is an obstruction to the microwave energy that slows down the velocity of the signal from the ideal velocity in air. The more dense the dielectric, the slower the energy moves.

The term *dielectric constant* (sometimes referred to as simply the DK or the *permittivity* of a material) goes right along with the definition of dielectric. The dielectric constant describes a material's density relationship to air. The dielectric constant of air is 1, so every other material has a dielectric constant greater than 1. Pure Teflon, for example, has a dielectric constant of 2.10 (listed as 2.08 in some tables). That means the velocity of the signal is decreased through Teflon to 69% of that in air. The velocity is decreased by the reciprocal of the square root of the dielectric constant. So, for Teflon, it is the square root of 2.1 (1.4491) divided into 1, or 0.69, or 69%. Similarly, a dielectric constant of 3.5 decreases the velocity of the signal to 53.4% (1/1.87) that in air, and a dielectric constant of 10.2 decreases the velocity 31.3% (1/3.1937) of the velocity in air.

As a rule of thumb, you should use low-dielectric materials at the higher end of the microwave spectrum. Doing that results in a much more reasonable length for transmission lines at the high frequencies, for example, when circuits with quarter-wave transmission line lengths have to be built. Materials with higher dielectric constants work better at the low end. These guidelines are for size considerations. At low frequencies, the wavelengths get considerably larger. By using a higher dielectric constant material, the lengths are easier to handle and put on a circuit board to maintain an architecture that will be small and conform to the

RF and microwave size constraints. These are very general rules, how-ever, since there is a tremendous overlap of material dielectric constants for different frequency ranges, as shown in Figure 6.1. It can be seen in the figure that usually more than one dielectric constant material can be used for a particular frequency range. For an application that operates from 1 to 2 GHz, for example, if you only looked at Figure 6.1, you could choose dielectric constants ranging from 2.32 to 10.2. That could get confusing and cause problems. For that reason, designers use other parameters to make their final choice.

One of the parameters that may determine which material to use is the *dissipation factor* (sometimes referred to as simply the DF). The dissipation factor is the loss in the material to RF and microwave energy. Another designation for this term is tan δ, which is the angle of a curve that indicates loss and conduction (Figure 6.2). The higher the conduction of the energy, the lower the angle and the smaller tan δ. That means the lower the value of dissipation factor (tan δ), the lower the losses and the better the material. Figure 6.2 tells us that for a small angle of δ to occur, we need a large conduction figure. That is understandable because the better the conduction in a circuit, the lower the losses are in that circuit.

Unlike the dielectric constant, the dissipation factor is a parameter that changes with frequency. The dielectric constant has some minor

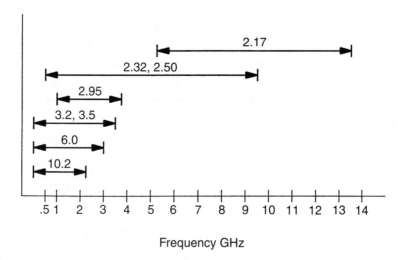

Figure 6.1   Dielectric constant chart.

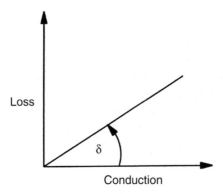

Figure 6.2 Dissipation factor.

variations over a large frequency range, but generally it is thought of as being constant and not a function of frequency. The dissipation factor is a loss and thus a function of frequency. As the frequency increases, so does the dissipation factor. That change is not linear, however; it usually is fairly low at low frequencies and as the frequency gets higher and higher it increases at ever greater rates until it approaches an exponential increase. Typical values for dissipation factor for RF and microwave materials are in the range of 0.004 to 0.006. These values are for frequencies in the area of 10 GHz. For the wireless frequency ranges, the values should be in the neighborhood of 0.003 to 0.0045. Such material would be considered to be low loss material and would find many applications in the RF, microwave, and wireless world.

The dissipation factor is used to find out if a filter will have a high insertion loss, if a directional coupler will have the proper coupling loss or insertion loss, or if a long delay line will have the proper loss or an extremely high loss that will prevent it from being used. Often, to check out a dissipation factor parameter for a material, bandpass filters are designed to operate at a variety of frequencies. When the filters are built and tested, the data show if the proper insertion loss and passband are present according to the initial design. If the loss is greater or the passband is smaller, the dissipation factor is not what the designers thought it was. At that point, they start to ask a few questions or look for another material for the task.

The next parameter was referred to in Chapter 3 in the description of stripline and microstrip transmission lines. That parameter is the *dielectric thickness*, also called the *b* dimension of the material, which is how it is designated in design equations. Recall from our discussions of stripline and microstrip that the dielectric thickness is a dimension of the material itself, not of the copper attached to it. However, since only one piece of material is used in a microstrip construction, the *b* dimension is the actual thickness of the material. The *b* dimension for stripline is actually twice the dimension of the material, because two pieces of material are used. Regardless, the dielectric thickness is really the thickness of only the material. Figure 6.3 shows the relationship between dielectric thickness and *b* dimensions for both stripline and microstrip. For stripline, the *b* dimension is used in design equations to determine the width of transmission lines or the spacing between transmission lines. That *b* dimension is two thicknesses of material, as shown in Figure 6.3(a). For microstrip, there is only one piece of material, so the *b* dimension and the dielectric thickness are the same number.

The common dielectric thicknesses are 0.020 and 0.030 in, and most manufacturers list their materials as being available from .005 to .060 in. The actual thickness, however, is not the most important number to look at when choosing the thickness of a material.

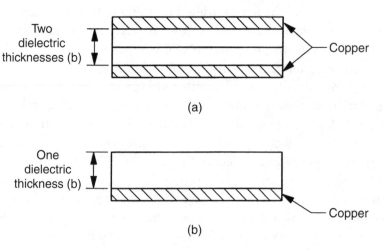

Figure 6.3 Dielectric thickness: (a) stripline and (b) microstrip.

The tolerance of the material is significantly more important. It is necessary to have very close tolerance on the dielectric thickness because it is so important in determining other parameters for components, such as characteristic impedance and coupling. Many of these tolerances are ±0.001, ±0.0015, or ±0.002 in, which are very close tolerance materials. We get further into the tolerances required for these materials as we progress into the chapter. For now, it is sufficient to say that the typical printed circuit board materials usually do not have such close tolerances and do not work well for RF and microwave applications. That is not to say those are bad materials. They are excellent for what they were designed to do and they do it very well. They just were not designed for RF and microwave work.

The next term to be investigated is *peel strength*. To understand this term, you must picture the construction of a piece of microwave material. For most cases, the construction is as shown in Figure 6.4, in which a dielectric material has copper attached to both sides of it. There are also cases in which there is copper on only one side of the dielectric. Regardless of the number of sides that have copper attached to them, the peel strength comes into play in all cases.

Peel strength is the amount of force required to separate the copper from the dielectric material. It is expressed in pounds per inch. This important parameter should be as high as possible to ensure that the copper stays attached to the dielectric material when it is manufactured, after it is sold, during fabrication, and when in use in a circuit or system. The last thing designers want to happen is to have their beautiful circuit literally fall apart when it gets out in the field. Some typical numbers for peel strength are 8, 10, 12, and 15 lb/in.

The copper must remain attached to the material during the fabrication process, and during the cutting and etching of the material itself.

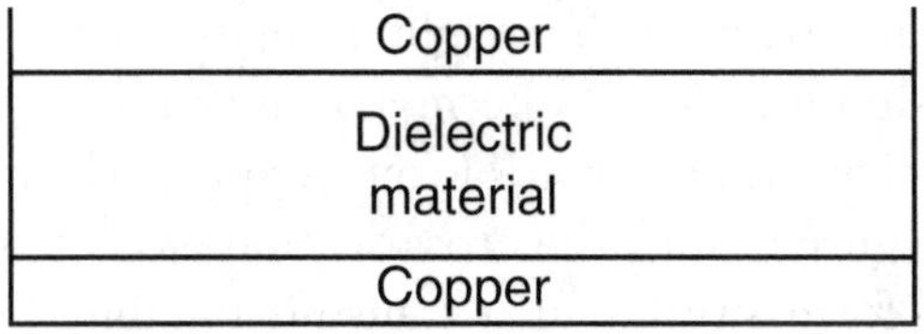

Copper
Dielectric material
Copper

Figure 6.4 Material construction.

Another area of fabrication that requires a certain minimum peel strength for a material is soldering. When a soldering iron is applied to a piece of copper attached to a dielectric material, that copper must remain attached as securely as it was before the iron was applied. If the peel strength of the material is too low, the copper will be dislodged from the material and the circuit will not operate properly.

It might seem questionable that the term *copper weight* is related to the thickness of copper. Actually, the copper weight specified on a materials data sheet is the weight of 1 sq ft of that copper. If we have standard 1-sq-ft pieces of copper and the only thing that can be changed from piece to piece is the thickness, then there is a very good connection between weight and thickness.

You will see copper weight designated as 1 oz, 2 oz, ½ oz, and so on. When we specify a material with 1-oz copper on it, we are saying that the copper is 0.0014 in thick. That is the standard copper on many RF and microwave materials. With that as the starting point, we can say that 2-oz copper is 0.0028 in thick, ½ oz is 0.0007 in thick, and ¼ oz is 0.00035 in thick. So it can be seen that there is a close relationship between the copper weight specification and the thickness of the copper on the RF and microwave materials.

Why are there so many different thicknesses of copper on these materials? If there is a standard thickness, why not use it for all circuit applications? The answer is the same as it has been throughout this text. A variety of applications come up, and no one material or copper-thickness requirement will take care of all of them. Thus, we have a variety of thicknesses from which to choose. If, for example, we are going to have some very narrow transmission lines in our circuit or are going to require some very small spaces between transmission lines, we probably would use a material with ½-oz copper on it so we would not have to etch off as much copper and thus would be able to get much better registration of lines with much straighter edges. Also, the cross-section of the transmission lines would be much more rectangular than if you had a longer etching time for a thicker copper, which can be seen in Figure 6.5. In Figure 6.5(a), the cross-section has nice steep sides on it and a flat top to form a rectangular transmission line. Figure 6.5(b) is a more typical cross-section of a transmission line. In that figure, there is a certain amount of undercutting, so the shape is not the ideal rectangle

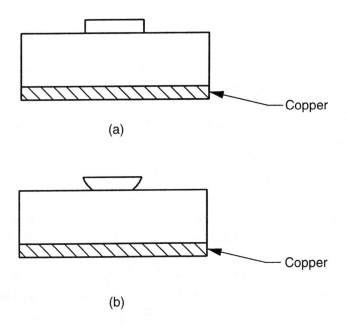

Figure 6.5 Transmission line cross-sections: (a) ideal cross-section and (b) actual cross-section with the proper copper weight.

but can be formed close to it if the etch time and the copper weight are taken into consideration. It can be seen what may happen if the copper weight is too large or small. If it is too large, it will take a long time to etch the copper, and the transmission line cross-section will be distorted. If the copper is too small, the copper will be completely undercut, and the transmission line will lift off the material. Neither condition is desirable for a properly etched transmission line in the RF and microwave spectrum.

With the copper weight defined, we now look at how the copper for the materials is fabricated. There are two methods of fabrication: *electrodeposited* (ED) and *rolled*. The standard method used for many materials is the ED copper. Rolled copper is usually a special order used for more critical RF and microwave applications.

If you look up *electrodeposition* in the dictionary, the basic definition says that it is the process of depositing a substance on an electrode by electrolysis. Unless you are a chemist, that definition probably tells you no more that you knew when you first looked up the term. You could look up the term *electrolysis*, which the dictionary says is the process of changing the chemical composition of a material by sending an electric current through it. If you put these two definitions together, you get some idea of what we are talking about. Basically we can say that electrodeposited copper is a material produced by a chemical process in which the individual copper particles are electrically joined to form the desired sheet thickness.

The process is similar to taking small pieces of clay and building one continuous sheet from the small pieces. If you had a certain number of pieces of clay placed next to one another over some base surface (as ED copper is plated on a rotating drum structure and then pulled off), you would have a structure that was a certain thickness and consistency. That can be likened to plating (or depositing) a layer of ½-oz copper (0.0007 in). If you need 1-oz copper (0.0014 in), you would repeat the process until the desired thickness was achieved. The electrodeposition process can be precisely controlled by controlling the time and the current used.

The easiest way to picture rolled copper is to compare it to putting a garment through an old-time wringer washer. For the many people who have never seen a wringer washer, it is an arrangement that has two rollers, one on top of the other, that are geared to a motor so they turn in opposite directions and feed a material through them. The tension can be adjusted on the rollers so you can squeeze a lot of water or only a little out of the clothes. The rollers used to make rolled copper are similar to those in a wringer washer. The copper block to be fabricated is sent through the roller combination, which has a force applied to each roller. The force applied determines the final thickness of the copper. Following the compression process are various processes that can be used on the copper to obtain the desired consistency and hardness. For electrical applications, the copper should be relatively soft to increase the conductivity in the material. That process is used for the more critical RF and microwave applications because the rolled copper is a much more evenly distributed and consistent copper than ED copper. That is not to say that ED copper is not a good conductor, but when critical applications that require the absolute minimum of losses in the material are called out, rolled copper may be better for the application. Rolled copper exhibits a lower loss characteristic because it is a much more uniform structure than ED copper. It should be pointed out that rolled copper costs more than ED copper.

The final term that we will define, *anisotropy*, is connected with the dielectric constant of a material. When we discussed dielectrics and the dielectric constant, we referred to the dielectric properties and dielectric constant in the X-Y plane, that is, ε_{xy}. There is, however, also a dielectric constant value that characterizes the parameter in the Z plane: ε_z. Anisotropy is the comparison of the two dielectric constants. Ideally, the ratio of the two, $\varepsilon_{xy}/\varepsilon_z$ equals 1.0, that is, the dielectric constant of the material is exactly the same in the X-Y plane as it is in the Z plane. That, however, usually is not the case. There usually is a small difference between the two numbers so that the ratio obtained is somewhat greater than 1.0. The difference in dielectric constants comes about because fill materials (ceramic, glass, etc.) are added to a pure material (e.g., Teflon) to obtain dimensional stability. When that happens, the material no longer is completely consistent in all directions. The difference is the anisotropy of the material. Some data sheets do not even list anisotropy as a standard parameter for a material. It may be necessary to contact the manufacturer to find out what the missing dielectric constant is.

Anisotropy is closer to 1.0 for lower dielectric constant materials. Consider the addition of fill materials to pure materials. The less fill you add, the lower the dielectric constant and the closer the material comes to a consistent material that exhibits the same properties regardless of the orientation used. Typical values for anisotropy are in the range of 1.025 to 1.040 for some low-dielectric Teflon fiberglass materials to 1.090 to 1.150 for some higher-dielectric materials of the same configuration.

Anisotropy is a term that some designers never encounter, while others deal with it every day. That is because not every design must be concerned with the dielectric constant in every direction. The major concerns come about in microstrip and stripline circuits, where the difference in dielectric constant produces additional capacitances at the edges of the line. Those capacitances are called *fringing capacitances*. The type of circuits that have additional fringing capacitance are those that have resonators (such as the filters discussed in Chapter 4), circuits with narrow transmission lines, and directional couplers that rely on coupling from the edge of the line rather than the side of the line. In all those cases, the effects of fringing capacitance must be known and taken into account. If you have a material with a high anisotropy and do not know it, the circuit probably will not perform as expected. So when you are choosing a material for which the listed types of circuits are used, be sure to determine what the anisotropy is for that material.

6.2   Material requirements

Microwave materials are special types of materials. The definitions presented in Section 6.1 give some indication of how they are special. To really understand just how special they are, we should answer two questions: "Why is it necessary to use special materials?" and "What makes these materials special?" The answers to those questions are evidence that RF and microwave materials actually are integral parts of the circuits they are used for.

To answer the first question, we refer to the typical epoxy printed circuit board used for radio and television circuit boards and that basically provides a support for components and gets current from one point to another. The specifications for a typical type of this material are as follows:

- Dielectric constant: 4.0 to 4.6;

- Dissipation factor: 0.018 to 0.025;

- Thickness tolerance: 0.031 in ±0.004 in; 0.062 in ±0.006 in.

The point to be emphasized with this information is the tolerances for the material. First, there is a 15% possible change in the dielectric constant of the material. The wavelength of a signal is found by dividing the speed of light by the product of the frequency and the square root of the dielectric constant. The wavelength calculation is used to find the length of transmission lines, so we can create specific situations to make the RF or microwave circuit function properly. If the dielectric constant of the material varies all over the place, it is difficult to determine a specific length and be able to characterize the transmission line for that circuit. Therefore, we need a material with a very close tolerance on the dielectric constant over a wide range of frequencies. You can see how a material that has such a close tolerance for the dielectric constant will pin down the wavelength to a specific number. If we need a transmission line 0.135λ long, we know that if we make the transmission line that length it will work and have the circuit perform properly. That is one area where we can use special materials.

The second area to look at is the thickness tolerance for the material (the dielectric thickness). The thickness varies from 10 to 12%. This is the b dimension for microstrip and $b/2$ for stripline (even though we still use a b designation for the ground-plane thickness). You will recall that certain relationships determine the width of a transmission line or the spacing between transmission lines when coupling is desired. Those relationships are the w/b ratio and the s/b ratio. The w dimension, which is the width of the stripline or microstrip transmission line, determines the impedance of the transmission line. If the b dimension varies a great deal, the impedance of the transmission line also will vary. That is an unacceptable condition in RF and microwave circuits. The impedance of the transmission lines must remain constant for every circuit that is built. Imagine what the result would be if you needed a 50Ω transmission line and the thickness of the material varied so much that the actual impedance you measured was 60Ω. That would create a mismatch where you did not expect a mismatch and would cause further problems in other circuits that followed, as well as those ahead of the circuit, since there now would be reflections coming back.

Also consider what would happen if you need a directional coupler and determine that you need a gap, using the given value of dielectric thickness, of 0.010 in. If the dielectric thickness from sheet to sheet of material or from one side of a single sheet to the other varies in such a manner that the gap effectively is 0.018 in, the directional coupler would have considerably more loss through the coupling than you planned. Therefore, the tolerance on the dielectric thickness must be very small for RF and microwave circuits to perform properly.

To further understand why special materials are needed, consider the following data for a typical RF and microwave material that is used for many applications throughout the industry:

- Dielectric constant: 2.50 ±0.05;

- Dissipation factor: 0.0015;

• Thickness tolerance: 0.031 in ±0.0015 in; 0.062 in ±0.002 in.

It can be seen from the data that the tolerances are much tighter than those of conventional printed circuit board material. Some quick calculations show that the dielectric constant is held within 2%, compared to a 15% change for the conventional material. Also, the thickness is held within ±5% in one case and ±3% for the other. That compares to 10 to 12% for the conventional material. So you can see how much tighter the tolerances are held in microwave material than in conventional material. Tighter tolerances guarantee that when you calculate a wavelength or line width for a specific material, the calculated value will be exact and remain constant over the frequency range used.

One more observation should be made concerning the comparison of the two materials. The dissipation factor for the conventional material is listed as 0.018 to 0.025, while the dissipation factor for the microwave material is 0.0015. That is an order of magnitude better for the microwave material. Recall that the lower the number, the less the loss. So the microwave material is also much better as far as loss is concerned.

Now we should be in a position to answer the questions posed at the beginning of this section. Actually, the answers to both questions are closely related. First, we use special materials for RF and microwave circuits because these circuits depend heavily on a tightly held value of dielectric constant and dielectric thickness, and the microwave materials ensure that all lengths and widths will be exactly as designed. Second, the RF and microwave materials are special because they do hold the tolerances as tightly as required. The dissipation factor is also much lower, which results in less loss through the material over the frequency of operation. Therefore, the two answers rely on each other and result in a closely controlled material that is ideal for RF and microwave applications.

6.3 Types of materials

We are now at a point where we can talk about some types of materials used for RF and microwave applications. The discussions here will be general in nature because of the great changes that are taking place in the materials market. Each type of material is presented with some typical specifications so that you can get an idea as to what each material is capable of doing. From the discussions, you should be able to make an educated decision as to which material is the best for a particular application.

6.3.1   Teflon fiberglass materials

A standard material for RF and microwave applications for many years was Teflon-based material. It was, and still is, an excellent choice for high-frequency circuit board material because of its uniform construction. Recall that the high-frequency energy propagates through the material in order to perform certain functions. If there are obstructions in the material, the energy slows down and is not as efficient. Teflon material allows an unobstructed trip through the material. As good as the Teflon material is for the electrical portion of the RF and microwave applications, it is almost as bad for the mechanical portion of those same applications. Teflon is a very soft material that will "cold flow" if pressure is applied to it. That means the material changes its thickness dimension under pressure and does not always come back to its original dimension when the pressure is removed. That is a problem because the dielectric thickness must remain in tact and be held to very close tolerances to have the circuits perform as expected. So, mechanically, pure Teflon is not one of the more acceptable materials to use. The key word here is *pure*. Most of the Teflon applications in RF and microwaves are not pure Teflon. It is a Teflon-based system with fillers to make the material more acceptable mechanically. The fillers usually are fiberglass, although recent advances use other materials.

The first Teflon-based material we will look at is the Teflon fiberglass material that was the first material to be used for RF and microwave applications. This material is usually referred to as PTFE-glass material. (PTFE is the abbreviation for *polytetrafluorethylene*, the chemical name for Teflon.) The fiberglass is used to reinforce the Teflon and make it more structurally rigid. It does that job very nicely, but it also changes the characteristics of the original Teflon material. We had a pure material (Teflon) and added an impurity (fiberglass) to it. That changes some of the parameters of the original material. Adding fiberglass to pure Teflon increases the dielectric constant of the material, but it also increases the losses. The dielectric constant of pure Teflon is 2.10 (although some reference books list it as 2.08). Adding fiberglass to Teflon raises the dielectric constant to approximately 2.60 before the losses become too large; the addition of more fiberglass only makes the performance worse.

Two methods are used to place the fiberglass in a Teflon material: woven and nonwoven. The woven type of structure looks like magnified cloth fabric. Figure 6.6 shows a basic structure for such a material. A base of Teflon has fiberglass strands woven throughout the material to form a very nice structure that performs well in RF and microwave applications. Figure 6.7 is a close-up picture of woven teflon fiberglass material.

As previously mentioned, the parameters are changed when we add fiberglass to a pure Teflon base. For an illustration, consider Table 6.1, which lists three woven PTFE-glass materials and three important parameters: dielectric constant, dissipation factor, and peel strength.

It can be seen from Table 6.1 that the parameters will change as more fiberglass is added. Material 1 has the lowest amount of fiberglass and material 3 the most fiberglass added to it. With that in mind, you can see that the dielectric constant has indeed increased with the amount of fiberglass, but the loss also has gone up. Another point to make here is that the peel strength increased when fiberglass was added to the pure Teflon. Remember that Teflon is also the material that keeps eggs from sticking to a frying pan. It is not too hard to see that the less effect the Teflon has on the material, the better the copper will adhere to the material. That can be seen in the table as an increase in the peel strength

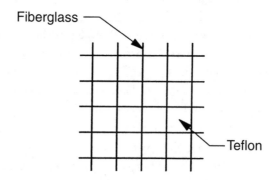

Figure 6.6   Woven PTFE structure.

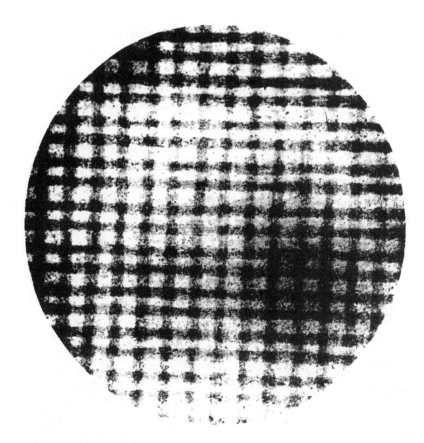

Figure 6.7 Woven Teflon material.

Table 6.1
Comparison of Parameters for Three Woven PTFE-Glass Materials

Parameter	Material 1	Material 2	Material 3
Dielectric constant	2.17 ±0.02	2.33 ±0.04	2.50 ±0.04
Dissipation factor	0.0009	0.0015	0.0018
Peel strength	8 lb/in	8 lb/in	12 lb/in

when the dielectric constant reaches 2.50. So the amount of fiberglass has a significant effect on many parameters for the RF and microwave materials.

The second type of PTFE-glass is nonwoven material, also called microfiber. In this Teflon-based material, the fiberglass particles are added in a random fashion and not the woven structure we saw in Figure 6.6. This type of structure produces a material with a dielectric constant variation of less than ±1% and low dissipation factors. It also is very stable, dimensionally being able to withstand temperatures up to 550°F (288°C) without warping. Figure 6.8 is a close-up picture of nonwoven Teflon fiberglass material.

Figure 6.8   Nonwoven Teflon material.

If we look at laminates (materials) with various amounts of fiberglass in a nonwoven structure and use the same parameters as in Table 6.1, we have the data given in Table 6.2 for nonwoven materials.

It can be seen from Table 6.2 that the same pattern is realized as in Table 6.1 with the woven material. As the fiberglass content increases, both the dielectric constant and the dissipation factor increase. The main difference between the two tables is that in Table 6.2 the peel strength is good for both materials and does not increase or decrease with the fiberglass content.

As stated previously, the introduction of fiberglass into a Teflon base increases the dielectric constant of the resultant material. Also, as the fiberglass content increases, the dissipation factor increases to a point where the losses are very high above a dielectric constant of about 2.60.

To allow the continued use of Teflon for higher dielectric constants, the filler must be changed. The material used as a filler for higher dielectric constants with reasonable dissipation factors is one that finds many applications throughout the field of RF and microwaves. That material is ceramic. With ceramic-filled PTFE, it is possible to obtain dielectric constants that range from 3.0 to 6.0 to 10.2. These materials find many applications for both high and low frequencies. To show how ceramic-filled PTFE materials compare to one another and to woven and nonwoven PTFE-glass materials, refer to Table 6.3. This table shows the dielectric constants of the available standard materials for many commercial RF and microwave applications.

There is one parameter shown in Table 6.3 that we have not yet discussed. The *coefficient of thermal expansion* is the amount that the material changes mechanically with temperature. This parameter, ex-

Table 6.2
Comparison of Parameters for Two PTFE Nonwoven Materials

Parameter	Material 1	Material 2
Dielectric constant	2.20 ±0.02	2.33 ±0.02
Dissipation factor	0.0009	0.0012
Peel strength	15 lb/in	15 lb/in

Table 6.3
Dielectric Constants for RF and Microwave Materials

Parameter	Material 1	Material 2	Material 3
Dielectric constant	3.0 ±0.04	6.0 ±0.15	10.2 ±0.30
Dissipation factor	0.0013	0.0025	0.0035
Peel strength	6–8 lb/in	6–8 lb/in	6–8 lb/in
Coefficient of thermal expansion	20–25 ppm/°C	20–25 ppm/°C	20–25 ppm/°C

pressed as parts per million per degree Centigrade, is a significant number for ceramic-filled materials. To understand just how important this parameter is, consider that the coefficient of thermal expansion for aluminum is 28 ppm/°C. If we can have a material that has basically the same coefficient of thermal expansion as the case it is placed in, we stand a much better chance of the circuit operating and remaining mechanically in tact over a wide range of temperatures. (The coefficient of thermal expansion for PTFE-glass material is 130 ppm/°C.) It can be seen that there is a large advantage in using a material such as the ceramic-filled PTFE materials.

6.3.2 Non-PTFE materials

For many years, the only materials available for RF and microwave applications were those with a Teflon base, the PTFE laminates. Those materials worked well for a long time and did a variety of tasks. With the change in markets in recent years, there has been the birth of a new line of materials that do not use Teflon but are reinforced hydrocarbon-ceramic laminates. Hydrocarbons are defined as any compound that contains only hydrogen and carbon. Materials such as benzene and methane are common hydrocarbons. This different combination has found many applications, especially since the common ceramic material is also used.

An important property of hydrocarbon-ceramic materials makes them more appealing than PTFE-based materials for some applications.

PTFE has one property that can cause some problems if used for certain designs. That property is a *transition region* at 19°C (68.2°F), which is shown as curve A in Figure 6.9. It can be seen from the figure that the dielectric constant is rather well behaved on both sides of the transition region, but the drop in dielectric constant is where the problem occurs. This natural property of Teflon has been tolerated over the years, since no other materials had been developed to eliminate the transition.

Curve B in Figure 6.9 is the curve for the hydrocarbon-ceramic material. It is a much more consistent curve, which allows the material to have a predictable dielectric constant over a wide range of temperatures. There is no sharp drop in the dielectric constant, so a designer can use some straightforward relationships to design a circuit that will perform as expected. One area where this type of material is valuable is in the design of filters. Recall the filter constructions shown in Chapter 4 with all the resonators and quarter-wave transmission lines used to build them. Their lengths depend heavily on the dielectric constant of the material used. If the dielectric constant varies over a temperature range, the filter response also varies and very well could be completely out of the range in which the filter was designed to operate.

Some properties of two hydrocarbon-ceramic materials are shown in Table 6.4.

The materials in Table 6.4 also have a coefficient of thermal expansion of 46 to 50 ppm/°C, which is very close to that of copper. When you

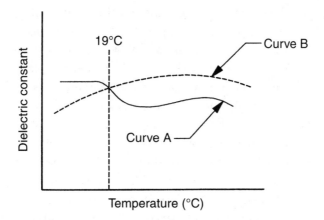

Figure 6.9 Dielectric constant vs. temperature.

Table 6.4
Properties for Two Hydrocarbon-Ceramic Materials

Parameter	Material 1	Material 2
Dielectric constant	3.38 ±0.05	3.48 ±0.05
Dissipation factor	0.002	0.004
Peel strength	6 lb/in	5 lb/in

consider this, you can understand how important it is to have the expansions of the copper and the material to which the copper is attached be basically the same. That makes for a very stable material over a wide temperature range.

6.3.3   Thermoset polymer composites

Another material that has come about from the increased commercial applications for RF and microwave circuits is the ceramic thermoset polymer composite. Once again, notice the use of ceramics for a high-frequency material. Ceramic comes up over and over again, and it certainly will continue to for many areas of RF and microwaves. These materials allow the material manufacturer to obtain a higher dielectric constant, good loss characteristics, and excellent material stability.

If you define all the terms, you get a better idea of what this material is. A thermoset is defined as a material that becomes permanently hard and unmoldable when once subjected to heat (certain plastics are thermosets); a polymer is a material that consists of many parts, either natural or synthetic; and a composite is defined as a material made up of many distinct parts. So what we have is a group of thermally combined materials with a ceramic base that will not soften when heated. This is an important property to have when a circuit is to operate in temperature extremes. The actual materials used in this type of laminate are, of course, proprietary, but they obviously are put together properly since they work so well. Table 6.5 lists some of the parameters for thermoset materials.

Table 6.5

Parameters for Three Thermoset Polymer Composites

Parameter	Material 1	Material 2	Material 3
Dielectric constant	3.27 ±0.016	6.00 ±0.080	9.80 ±0.24
Dissipation factor	0.0016	0.0018	0.0015
Peel strength	3 lb/in	3 lb/in	3 lb/in

6.3.4 Polyester materials

One of the most recent materials to come on the scene, as of this publication, might seem unlikely to be associated with RF and microwave circuits. This material is a polyester (yes, the material your clothes are made of). Once again, there is a proprietary process that makes this material compatible for use in RF and microwave circuits, but it also makes a nice high-frequency material that is finding its place in the industry for many applications. Table 6.6 shows the critical parameters for the polyester material.

It can be seen that the polyester material has the parameters to be a viable consideration for many RF and applications. Many areas of filters, antennas, couplers, and amplifiers already have found the polyester material to their liking.

Table 6.6

Parameters for Polyester Material

Parameter	Material
Dielectric constant	3.20 ±0.04
Dissipation factor	0.0025

6.4 Choice of materials

To answer the question of how to choose an RF and microwave materials, one must know many factors about the application. It is not a question that can be answered by always saying material A is best or material B is ideal. Many factors come into play. Many different types of materials are available, and no one material will do every task. So, the choices must be studied before the final decision can be made. That is much different from years ago, when only one material was available. Many times that one material was not good for an application, but it had to be adapted since it was the only one you could get. Today's designers are faced with the opposite dilemma: many, many materials are available and they have to sort through all the claims and advertising to determine just which material will do the job.

In the choice of a material, two closely related factors usually come up first: the dielectric constant and frequency. Figure 6.1 related dielectric constants and frequency. In that figure were areas that overlapped, but generally you could begin to choose a material from the chart. The dielectric constant is important and related to frequency, because the lengths of the transmission lines will be determined by the dielectric constant of the material and their frequency of operation. What you should be looking for is a dielectric constant that will produce transmission lines of a "reasonable" length. For most applications, a transmission line that is supposed to be a quarter-wavelength should not be 25 in long or even be 0.010 in long. Those would be unreasonable lengths. Use some common sense and choose a material that will give you some good lengths.

The next parameter to looked at is the dissipation factor, which is the loss characteristic of the material. If the operation of a circuit depends on having very low losses in the circuit, choose low-dissipation materials for that application. If it is possible to compromise on this parameter, you probably can use a variety of materials for your circuit. Dissipation factor for filters and antennas should be one of the primary parameters to look at in the choice of a material.

Such mechanical properties as peel strength, coefficient of thermal expansion, and dielectric thickness all should be looked at. The peel strength ensures that the copper is going to stay in place when the circuits are in operation. The coefficient of thermal expansion should be looked at closely if the application is to be subjected to temperature extremes. Careful examination of this parameter and of the interfaces of materials, such as copper, aluminum, and the material, ensures that there will be no broken connections or removal of ground planes during the temperature extremes. The dielectric thickness determines the widths of the transmission lines you will be using and lets you know if they will be easy to bend around corners, if necessary. Also, the thickness lets you know if you will be able to construct, for example, a directional coupler with a very small gap between the main line and the coupled line.

Obviously there are other parameters on a material data sheet, but the ones discussed here are the important ones that should be considered for the majority of applications for RF and microwaves. It cannot be stated too strongly that it is important to really look at the materials that are available and choose the one that is going to do a given job. The answer to the question of how to choose an RF and microwave material is "very carefully." Remember, the best material that is available today is the one that will do the job.

6.5 Summary

This chapter covered an area that many times is overlooked or even ignored by many people. They still consider the RF and microwave material to be a *printed wire board* (PWB) and do not understand how important this material is.

The chapter began by defining many of the terms that relate to RF and microwave materials: dielectric, dielectric constant, dissipation factor, dielectric thickness, peel strength, copper weight, and anisotropy. This chapter also showed how those properties influence the operation of a circuit that is placed on the material.

Next, we got into material requirements, why we need special materials, and what makes them special. The next section described five types of materials: Teflon-fiberglass materials, ceramic-filled PTFE, non-PTFE materials, thermoset polymers, and polyester.

Finally, the critical choice of the material to use was covered. This process should take just about as much time as the design itself. It is that important. Some of the critical parameters were outlined and steps were shown that will help in the choice of the proper material.

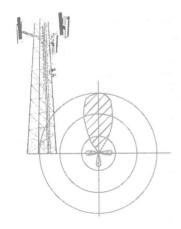

Glossary

amplitude balance   The comparison of the two power levels at the output of a quadrature hybrid. It indicates the difference in amplitude, if any, between the two outputs.

anisotropy   The ratio of the dielectric constant in the X-Y plane to the dielectric constant in the Z-plane ($\varepsilon_{xy}/\varepsilon_z$).

attenuate   To lessen the value of a signal. Attenuation is measured in decibels.

automatic leveling circuit (ALC)   A circuit that takes an unleveled RF input signal and provides an output that is flat across a wide band of frequencies.

average power   The peak power of a signal multiplied by the duty cycle of the system.

bandwidth   The band of frequencies over which a component or a system works; also, the band of frequencies over which the specifications of a data sheet are valid.

bipolar device   A device that has a construction such that there are majority carriers (electrons) and minority carriers (holes) moving through the device, for example, a bipolar transistor.

carrier lifetime   The measure of the ability of a PIN diode to store an electrical charge. It is expressed in nanoseconds and is improved by reducing impurities in the diode.

cellular telephone   A mobile telephone that relies on cell stations throughout the area to provide communications between parties.

characteristic impedance   A dynamic impedance, in ohms, that is constant throughout the transmission line. For RF and microwave applications, the characteristic impedance usually is 50Ω.

chips   Sections of the time and frequency domain that are used in spread spectrum secure communications systems. They are there for a specific period of time at a certain bandwidth of frequency.

coaxial cable   A high-frequency transmission line with a center conductor surrounded by a dielectric and a ground shield. It is basically one line inside another.

code division multiple access (CDMA)   A scheme used in spread spectrum secure communications with chips at specific times and frequencies.

coefficient of thermal expansion   The amount that a material changes with temperature, expressed in parts per million per degree centigrade.

conversion loss   The loss through a mixer that is measured from the RF input to the LO output.

coplanar waveguide   A transmission line where distributed element circuits are used and the circuit and the ground plane are on the top of the circuit board. A special case is the structure described with a ground plane also on the bottom of the circuit board (ground-backed coplanar waveguide).

copper weight   The designation used to describe the thickness of the copper on a microwave circuit board. It is the weight of 1 ft^2 of copper.

coupler   A component that moves energy from one circuit to another without any direct connection. The energy is "coupled" between circuits by the electric field that is set up by the signal traveling down the line.

cutoff frequency   The frequency in a filter (lowpass and highpass) where the response drops to half power (3 dB).

decibel (dB)   A relative number that is the ratio of two powers or two voltages. There is no unit associated with dB.

decibels referred to milliwatts (dBm)   An absolute power number that is the ratio of a power compared to 1 mW (0 dBm is 1 mW).

dielectric   An obstruction to RF and microwave energy that slows down the velocity of the signal from the ideal velocity in air.

dielectric constant   The value of an RF and microwave material that tells you its density in relation to air.

dielectric thickness   The dimension that measures the distance between the copper on both sides of the material. It is the actual thickness of the dielectric material.

directional detector   A component that consists of a directional coupler and detector built into one unit.

directivity   A parameter of a directional coupler that is determined by taking the value of the isolation of the coupler and subtracting the coupling. It is a measure of the desired signal strength to the undesired signal strength.

dissipation factor   The RF and microwave loss in a material. It may be referred to as the DF or the tan δ.

distributed element component   A component that is distributed over the surface of a high-frequency circuit board. This is much more efficient since it takes advantage of the skin effect at high frequencies.

Doppler effect   The change in frequency of a signal as it approaches or leaves a reference point.

downconverter   A circuit that takes an incoming signal and converts it to a lower-frequency output. This is accomplished in a superhetrodyne circuit, in which the mixer produces an intermediate frequency that is lower than the incoming RF signal.

duty cycle   The amount of time that a pulse is actually on in a pulsed system. It is the ratio of the pulse width to the pulse repetition rate.

effective dielectric constant   The dielectric constant used for microstrip calculations. It is the result of the combination of the dielectric constant of the material and the dielectric constant of the air that is at the top of the material.

effective isotropic radiated power (EIRP)   The actual power radiated from the antenna. It takes into account the power amplifier output and the antenna gain.

electrodeposited (ED) copper   A material produced by a chemical process in which the individual copper particles are electrically joined to form the desired sheet thickness.

epitaxial   A layer formed by the condensation of a single crystal film of semiconductor material on a wafer. An epitaxial device is one in which the collector region is formed on a low-resistivity silicon substrate.

far field   The area beyond the near field, with the near field being that field generated very close to an antenna; also called the radiation field.

field effect device   A device that is also called a unipolar device since there are only majority carriers present within the device. The common device is the field effect transistor (FET).

frequency division multiple access (FDMA)   A scheme in which the frequency is divided into bands and the signal is on all the time.

gigahertz (GHz)   A unit of frequency measurement that is 1×10^9 Hz (cycles per second), or 1 billion Hz.

ground-plane spacing   In stripline, the spacing of the two dielectric materials or the distance between the two copper layers, or ground planes.

guard band   A small band of frequencies placed between the desired bands of frequencies in the FDMA scheme.

guard time   A small period of time between the allocated time slots in TDMA.

gyromagnetic motion   A circular motion caused by the interaction of a magnetic field and a ferrite device.

HEMT   High electron mobility transistor.

insertion loss   The low-loss parameter of RF and microwave components. This is the loss through the straight-through port of a coupler or the loss in the passband of a filter.

intrinsic   A perfect, or pure, material. Usually it has either no impurities in it or very few.

lumped circuits   Standard resistors, capacitors, and inductors that have their entire values lumped into the components and not distributed over a larger area.

major lobe   The area of a radiation pattern for an antenna where the largest majority of the energy is concentrated.

megahertz   A unit of frequency measurement that is 1×10^6 Hz (cycles per second), or 1 million Hz.

MESFET   Metal-semiconductor field effect transistor.

microstrip   An RF and microwave transmission line that is made up of a dielectric material with a circuit etched on one side. The top of the circuit is air.

microwaves   A radio wave operating in the frequency range of 500 MHz to 20 GHz that requires printed circuit components instead of conventional lumped components.

minimum negative resistance The smallest value of negative resistance in a tunnel diode.

minor lobe A secondary, or lower-level, lobe of an antenna radiation pattern. Side lobes and the back lobe are secondary lobes.

mobile telephone switching office (MTSO) An integral part of the cellular telephone system.

near field The close-in area of an antenna; also called the induction field.

passband The band of frequencies in which the loss is at a minimum value.

peak power In a pulse system, the amount of power present at the top of the pulse. This is as opposed to the continuous wave (CW) power, which is on all the time.

peel strength The amount of force, in pounds per inch, required to remove the copper from a dielectric material.

permittivity Another way of designating the dielectric constant of a material.

personal communications network (PCN) Mobile communications network that interfaces with the standard telephone system.

phase balance The difference in phase between the two output ports of a quadrature hybrid.

PN junction A junction of two semiconductor materials. One has a doping that makes it basically positive, and the other is basically negative.

polytetrafluorethylene (PTFE) The chemical term for Teflon.

pulse repetition rate The amount of time between pulses in a pulsed system; usually designated as T.

pulse width In a pulsed system, the width of the pulse in seconds; usually designated as τ.

quadrature hybrid A coupler that provides equal amplitude output levels that are 90 degrees apart in phase.

radiation pattern A pattern of gain versus angle that characterizes the radiation of an antenna. It contains major and minor lobes that show what area is best for operation of that particular antenna.

reflection coefficient The percentage of signal that is being reflected back from a mismatch.

return loss The level of signal that is being reflected back from a mismatch condition; expressed in decibels.

ripple The amplitude variation of a filter response. This is in the passband of the filter.

rolled copper A method of producing the desired thickness of copper for a material. It is produced by running the copper through a set of rollers that are under pressure.

Schottky junction The junction of a semiconductor and a metal, usually aluminum. This type of device is used for high-frequency diodes and transistors.

semirigid cable A coaxial cable with a solid-center conductor, solid dielectric, and solid outer conductor.

short circuit A condition in which the center conductor of a transmission line is connected directly to ground. This arrangement is used to establish a reference for impedance measurements.

skin effect The condition in which the high-frequency current in a transmission line travels only on the "skin" of the line.

standing wave A wave that results when there is a mismatch at the end of the transmission line. It is the combination of the forward and the reflected waves and appears to stand still on the line.

stripline A transmission line that consists of two pieces of material with the circuit sandwiched between them. There is a ground plane on both sides of this structure.

tangential sensitivity (TSS) A measure of the lowest signal that can be detected by a microwave detector.

thermal resistance The resistance present in a power transistor that tells how well the heat will be taken away from the chip inside the device.

time division multiple access (TDMA) A process used for many digital communication applications, in which signal transmissions are time shared and not on all the time.

transition region The area around a PN junction.

transmission line A device used to transmit energy from one point to another efficiently.

upconverter A system that uses a mixer to take an incoming signal and change it to a higher output frequency.

voltage standing wave ratio (VSWR) The voltage representation of the standing wave.

wavelength The distance on a signal between corresponding points on that signal: minimum to minimum, maximum to maximum, and so on.

wireless A communications system that accomplishes its function of transmission from one point to another without the aid of cables or wires.

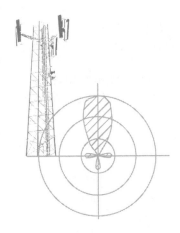

About the Author

THOMAS S. LAVERGHETTA is an associate professor at Purdue University's Fort Wayne campus. He teaches all the electronic communications courses at Purdue, including an introductory microwave course.

Mr. Laverghetta spent 23 years in industry prior to teaching full time. He was a design engineer at ITT Aerospace, Magnavox, Anaren Microwave, and General Electric Heavy Military Electronic Systems. Prior to receiving his BSEE from Syracuse University, he was a microwave technician at the Syracuse University Research Corporation and at General Electric.

Mr. Laverghetta has designed stripline and microstrip circuitry for many space applications, shipboard and ground equipment, airborne equipment, and test stations.

He is the author of *Microwave Measurements and Techniques, Handbook of Microwave Testing, Microwave Materials and Fabrication Techniques,* first and second editions, *Practical Microwaves, Modern Microwave Measurements and*

Techniques, Solid State Microwave Devices, Analog Communications for Technology, and *Practical Microwaves.* He also has authored many papers for journals and conference proceedings.

Mr. Laverghetta received his MSEE from Purdue University, is a Senior Member of IEEE and a member of ASEE (American Society for Engineering Education), and is an Accredited Professional Consultant through the American Consultants League.

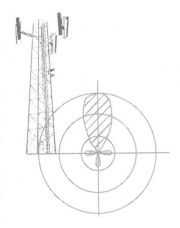

Index

The Artech House Microwave Library

Low Phase Noise Microwave Oscillator Design, Robert G. Rogers

MATCHNET: Microwave Matching Networks Synthesis,
Stephen V. Sussman-Fort

Matrix Parameters for Multiconductor Transmission Lines: Software and User's Manual, A. R. Djordjevic et al.

Microelectronic Reliability, Volume I: Reliability, Test, and Diagnostics,
Edward B. Hakim, editor

Microelectronic Reliability, Volume II: Integrity Assessment and Assurance, Emiliano Pollino, editor

Microstrip Lines and Slotlines, Second Edition, K.C. Gupta,
Ramesh Garg, Inder Bahl, and Prakash Bhartia

Microwave and Millimeter-Wave Diode Frequency Multipliers,
Marek T. Faber, Jerzy Chamiec, Miroslaw E. Adamski

Microwave and RF Circuits: Analysis, Synthesis, and Design,
Max Medley

Microwave and RF Component and Subsystem Manufacturing Technology, Heriot-Watt University

Microwaves and Wireless Simplified, Thomas S. Laverghetta

Microwave Circulator Design, Douglas K. Linkhart

Microwave Engineers' Handbook, Two Volumes, Theodore Saad, editor

Microwave Materials and Fabrication Techniques, Second Edition,
Thomas S. Laverghetta

Microwave MESFETs and HEMTs, J. Michael Golio *et al.*

Microwave and Millimeter Wave Heterostructure Transistors and Applicatons, F. Ali, editor

Microwave and Millimeter Wave Phase Shifters, Volume I: Dielectric and Ferrite Phase Shifters, S. Koul and B. Bhat

Microwave and Millimeter Wave Phase Shifters, Volume II: Semiconductor and Delay Line Phase Shifters, S. Koul and B. Bhat

Microwave Mixers, Second Edition, Stephen Maas

Microwave Transmission Design Data, Theodore Moreno

Microwave Transition Design, Jamal S. Izadian and Shahin M. Izadian

Microwave Transmission Line Couplers, J. A. G. Malherbe

Microwave Tubes, A. S. Gilmour, Jr.

Microwaves: Industrial, Scientific, and Medical Applications, J. Thuery

Microwaves Made Simple: Principles and Applicatons,
Stephen W. Cheung, Frederick H. Levien et al.

MMIC Design: GaAs FETs and HEMTs, Peter H. Ladbrooke

Modern GaAs Processing Techniques, Ralph Williams

Modern Microwave Measurements and Techniques,
Thomas S. Laverghetta

Monolithic Microwave Integrated Circuits: Technology and Design,
Ravender Goyal et al.

*MULTLIN for Windows: Circuit-Analysis Models for Multiconductor
Transmssion Lines, Software and User's Manual*,
Antonije R. Djordjevic, Darko D. Cvetkovic, Goran M. Cujic,
Tapan K. Sarkar, Miodrag B. Bazdar

Nonuniform Line Microstrip Directional Couplers, Sener Uysal

PC Filter: Electronic Filter Design Software and User's Guide,
Michael G. Ellis, Sr.

*PLL: Linear Phase-Locked Loop Control Systems Analysis Software and
User's Manual*, Eric L. Unruh

RF Design Guide: Systems, Circuits, and Equations, Peter Vizmuller

*Scattering Parameters of Microwave Networks with Multiconductor
Transmission Lines: Software & User's Manual*,
A. R. Djordjevic et al.

Solid-State Microwave Power Oscillator Design, Eric Holzman and
Ralston Robertson

Terrestrial Digital Microwave Communications, Ferdo Ivanek et al.

Transmission Line Design Handbook, Brian C. Waddell

*TRAVIS Pro: Transmission Line Visualization Software and User's
Manual, Professional Version*, Robert G. Kaires and
Barton T. Hickman

*TRAVIS Student: Transmission Line Visualization Software and User's
Manual, Student Version*, Robert G. Kaires and Barton T. Hickman

Yield and Reliability in Microwave Circuit and System Design,
Michael Meehan and John Purviance

For further information on these and other Artech House titles, including previously considered out-of-print books now available through our In-Print-Forever™ (IPF™) program, contact:

Artech House
685 Canton Street
Norwood, MA 02062
781-769-9750
Fax: 781-769-6334
Telex: 951-659
e-mail: artech@artech-house.com

Artech House
Portland House, Stag Place
London SW1E 5XA England
+44 (0) 171-973-8077
Fax: +44 (0) 171-630-0166
Telex: 951-659
e-mail: artech-uk@artech-house.com

Find us on the World Wide Web at:
www.artech-house.com

Learning Resources
Centre